Pei He

On the structure-property correlation and the evolution of Nanofeatures in 12–13.5% Cr oxide dispersion strengthened ferritic steels

**Schriftenreihe
des Instituts für Angewandte Materialien**
Band 31

Karlsruher Institut für Technologie (KIT)
Institut für Angewandte Materialien (IAM)

Eine Übersicht über alle bisher in dieser Schriftenreihe erschienenen Bände finden Sie am Ende des Buches.

On the structure-property correlation and the evolution of Nanofeatures in 12–13.5% Cr oxide dispersion strengthened ferritic steels

by
Pei He

Dissertation, Karlsruher Institut für Technologie (KIT)
Fakultät für Maschinenbau
Tag der mündlichen Prüfung: 11. November 2013

Impressum

Karlsruher Institut für Technologie (KIT)
KIT Scientific Publishing
Straße am Forum 2
D-76131 Karlsruhe

KIT Scientific Publishing is a registered trademark of Karlsruhe
Institute of Technology. Reprint using the book cover is not allowed.

www.ksp.kit.edu

Print on Demand 2014

ISSN 2192-9963
ISBN 978-3-7315-0141-1

On the structure-property correlation and the evolution of
Nanofeatures in 12–13.5% Cr oxide dispersion strengthened
ferritic steels

zur Erlangung des akademischen Grades eines

Doktors der Ingenieuwissenschaften

der Fakultät für Maschinenbau des

Karlsruhe Instituts für Technologie (KIT)

genehmigte Dissertation von

Pei He

aus Zhengzhou, China

Tag der mündlichen Prüfung: 11.11.2013

Hauptreferent: Prof. Dr. Hans Jürgen Seifert

Korreferent: Prof. Dr.-Ing. habil. Marc Kamlah

Acknowledgments

This work has been conducted at Institute for Applied Materials-Applied Material Physics (IAM-AWP) of Karlsruhe Institute of Technology (KIT). The financial support from both China Scholarship Council (CSC) and KIT is gratefully acknowledged. I would like to express my appreciation to all people who have contributed to this research both directly and indirectly, many of who are not included here.

I acknowledge greatly Professor Seifert for the scientific advice and careful review of the dissertation. I would like to thank the committee for being generously flexible in regard to scheduling of the defense times for both the 'proposal' of this work as well as the dissertation defense.

This work would never be possible without the contribution from my supervisors Dr. A. Möslang and Dipl.-Ing. R. Lindau. They provided me great support, guidance and encouragement on my work. They also helped to shape me and my opinions of scientific research.

Many people have been involved in this research through various collaborations. Special thanks to Dr. M. Klimenkov and Mrs. Jäntsch for the patient guide of sample preparation and TEM investigation. Mr. Ziegler offers me great help for the material fabrication. I would thank Mrs. Materna-Morris, Mr. Zimmermann, and Mr. Bolich for scanning electron microscopy and metallographic training. I would also express my appreciation to Mr. Baumgärtner for the mechanical tests and IT support. The roughness measurements of fatigue samples were conducted with the help of Mrs. Torte. Heart-felt thanks to Dr. Adelhelm and Dr. Bergfeldt for the chemical composition analysis. Dr. Leiste provided great assistance on XRD test. Many thanks go to Dr. Franke for the fruitful thermodynamics discussion.

I would specially acknowledge Dr. T. Liu (from ISS, KIT), Dr. Kurinskiy, J. Hoffmann, Dr. Commin for the kind participation in X-ray absorption fine-structure spectroscopy. In addition, timely assistance for EXAFS experiments from Dr. Nikitenko and Dr. Silveira at ESRF is much appreciated.

Prof. Sandim assisted EBSD measurement and provided many insightful discussions regarding the result.

I am very grateful to all the colleagues of our group, for warm working atmosphere, many beneficial works and private discussions, and support with solving all kinds of problems. I would like to specially thank my officemate, Dr. Kurinskiy for cheerful atmosphere in our office. I would specially thank Jan Hoffmann for the German translation of the abstract of this thesis.

I also want to acknowledge all my friends that made my stay in Karlsruhe so pleasant. In particular, I want to mention Yiming Wang, Yi Zhong, Ke Ren, Yi Feng, Yan Wang and Feng Xu for their warm friendship.

As for as the professional acknowledgements, first and foremost I want to acknowledge my boyfriend, Weizhi Yao, for his continuous love, support and encouragement. He is the best man for me to explore the wonderful world in both professional and personal ways. Finally, I want to thank my mother, my father and my brother for their endless love, support and patience during the last few years.

Abstract

Oxide-dispersion-strengthened (ODS) ferritic steels are primary candidate materials for structural applications in nuclear fusion reactors and fourth generation fission reactors due to their high temperature stability and strength, and resistance to radiation damage. The manufacturing of ODS ferritic alloys is based on the powder metallurgy route commonly involving mechanical alloying (MA) of the base powder with Y_2O_3, then consolidation by hot isostatic pressing (HIP). Appropriate thermo-mechanical processing (TMP) is often performed to improve the microstructure and the mechanical properties. Main objective of this Ph.D thesis is to develop, by systematic variation of the chemical composition, and TMP, 14% Cr nano-structured ferritic alloys with significantly improved high-temperature properties compared to currently available ODS alloys. Application of state-of-the-art characterization tools shall lead to an integrated understanding of structure-property correlation and the formation mechanism of nanoparticles.

Detailed TEM studies show a bimodal grain size distribution and pronounced Cr carbides along grain boundaries in the as-hipped ODS steels. Annealing treatment at 1050 °C leads to complete dissolution of carbides into the matrix and enhanced hardness. After TMP, heterogeneous microstructure in terms of elongated grain morphology and broad grain size range is revealed by electron backscattered diffraction (EBSD). The presence of a sharp α- (<110> // rolling direction (RD)) texture fiber and a large fraction of low angle grain boundaries indicate strong recovery.

The characteristics of nanoscale oxide particles have been investigated by means of TEM. Ti addition is not only beneficial to refine the particle size and increase the particle number density, it can lead to significant change in the chemical composition of ODS dispersoids. The addition of 0.3 wt. % Ti exhibits the best refinement effect of nanoscale ODS particles by reducing the average diameter from 11.48 to 3.73 nm. Moreover, a majority of Y_2O_3 is replaced by Y-Ti-O particles in all Ti-containing ODS alloys. Energy-dispersive X-ray (EDX) mapping provides reliable proof of the coexistence of Y_2O_3 and a small number of Y-Cr-O particles in Ti free ODS alloy. No core-shell structure is found in any of the analyzed nanoparticles.

TMP displays a positive effect on Charpy impact property with a slight increase in upper-shelf energy (USE) from 1.5 J to 2.1 J for 13.5Cr1.1W ODS steel. In comparison to EUROFER ODS, the insufficient improvement can be attributed to the occurrence of crack-divider type delamination due to the heterogenous microstructure. Substantial fatigue lifetime prolongation by a maximum factor of 20 and constant stress amplitude of over the entire fatigue life are achieved for 13.5Cr1.1W ODS steel after TMP irrespective of the applied strain range. The extraordinary fatigue resistance of ODS steels originates from stable grain structure, constant dislocation densities of 10^{14} m^{-2}, the absence of evident chemical segregation at grain boundaries and highly stabilized nanoscale oxide particles with an average diameter of about 4 nm.

X-ray absorption fine structure (XAFS) spectroscopy performed at European Synchrotron Radiation Facility (ESRF) provides a non-destructive way to study the physical and chemical state of dilute species in ODS alloys and thus enables a deeper understanding of the evolution of Y_2O_3 during the fabrication process. A linear combination fit of the X-ray absorption near-edge fine structure (XANES) suggests that only 10–14% of the initially added 0.3 wt. % Y_2O_3 dissolves into the steel matrix after 24 hours MA. Fitting of the extended X-ray absorption fine structure (EXAFS) indicates the coexistence of Y_2O_3 and $YCrO_3$ in the Ti-free compacted ODS sample which is validated by EDX mapping. The more dispersed Y-enriched oxides in the compacted ODS steels are a clear indication of the coexistence of crystalline nanoparticles and disordered phases after high temperature consolidation and annealing.

Kurzfassung

Oxid dispersionsgehärtete ferritische (ODS) Stähle sind auf Grund ihrer Hochtemperaturfestigkeit und Strahlungsresistenz geeignet für die Anwendung als Strukturwerkstoffe in Fusionsreaktoren und Generation IV Kernkraftwerken. Die Herstellung von ferritischen ODS Legierungen erfolgte durch einen pulvermetallurgischen Prozess: mechanisches Legieren eines vorlegierten Pulvers mit Y_2O_3 und anschließende Kompaktierung durch heiß-isostatisches Pressen (HIP). Geeignete thermo-mechanische Nachbehandlungen wurden durchgeführt, um die mikrostrukturellen und mechanischen Eigenschaften zu verbessern. Das Hauptziel dieser Dissertation ist die Verbesserung einer nano-strukturierten 14%Cr Legierung gegenüber zur Zeit erhältlichen ODS Legierungen im Hinblick auf deren Hochtemperatureigenschaften. Dies geschah durch systematische Variation der chemischen Zusammensetzung und thermo-mechanischen Nachbehandlung. Durch die Anwendung von hochmodernen Charakterisierungsmethoden sollte ein ganzheitlicheres Verständnis von Struktur-Eigenschafts Korrelationen und der Bildungsmechanismen der Nanopartikel erreicht werden.

Detaillierte TEM-Untersuchungen an den Legierungen nach dem HIP zeigen einen bi-modale Korngrößenverteilung und Chromkarbide, die an Korngrenzen segregieren. Durch eine Glühung bei 1050°C kann eine vollständige Auflösung dieser Karbide im Werkstoffgefüge und damit eine Steigerung der Härte erreicht werden. Nach der thermo-mechanischen Behandlung konnte mittels Electron Backscatter Diffraction (EBSD) eine inhomogene Mikrostruktur mit langgestreckten Körnern und einer breiten Korngrößenverteilung nachgewiesen werden. Das Vorhandensein einer scharfen α- (<110> // RD) Fasertextur und der große Anteil an Kleinwinkelkorngrenzen weisen auf starke Erholung hin.

Die nanoskaligen Oxidpartikel wurden mittels TEM untersucht. Die Zugabe von Titan bewirkt nicht nur eine Verkleinerung der Oxidpartikelgrößen und eine Erhöhung ihrer Anzahl, sondern auch eine signifikante Änderung der chemischen Zusammensetzung. Die Zugabe von 0.3 Gew.% Titan bewirkte die größte Verfeinerung, mit einer Reduktion des mittleren Durchmessers von

11.48 auf 3.73 nm. Zusätzlich wird ein Hauptteil der Y_2O_3-Partikel durch Y-Ti-O Partikel in allen titanhaltigen Legierungen ersetzt. Energiedispersive Röntgenspektroskopie (EDX) liefert einen verlässlichen Beweis für die Koexistenz von Y_2O_3 und einer kleinen Anzahl Y-Cr-O Partikel in den titanfreien Legierungen. Es wurden keine Kern-Schale-Strukturen bei den untersuchten Nanopartikeln gefunden.

Die thermo-mechanische Nachbehandlung hat einen positiven Effekt auf die Zähigkeit der Materialien im Kerbschlag-Biegeversuch, mit einer Erhöhung der Hochlagen-Energie von 1.5 J auf 2.1 J beim 13.5Cr1.1W ODS Stahl. Die im Vergleich zu EUROFER ODS ungenügende Verbesserung ist auf Riss-Verteilungs Delamination zurückzuführen, welche durch die inhomogene Mikrostruktur hervorgerufen wird. Beträchtliche Lebensdauerverlängerungen mit einem maximalen Faktor von 20 konnten für die 13.5Cr1.1W ODS Legierung bei Tests mit einer konstanten Spannungsamplitude über die gesamte Ermüdungsdauer erreicht werden. Diese außerordentliche Ermüdungsbeständigkeit der ODS Stähle wird durch die stabile Mikrostruktur und die konstante Versetzungsdichte von 10^{14} m^{-2} hervorgerufen. Dies wird unterstützt durch das Fehlen von Seigerungen an den Korngrenzen sowie durch hochstabile nanoskalige Oxidpartikel mit einer durchschnittlichen Größe von ca. 4 nm.

Röntgenabsorptions Feinstrukturspektroskopie (XAFS), die an der europäischen Synchrotonstrahlungsquelle (ESRF) in Grenoble durchgeführt wurde, bot einen zerstörungsfreien Weg, um die physikalischen und chemischen Zustände der Elemente, die in schwachen Konzentrationen vorliegen, zu bestimmen. Die gewonnenen Ergebnisse erlauben ein tieferes Verständnis der Entwicklung des Y_2O_3 während des Herstellungsprozesses. Eine Linearkombinationsanpassung der Röntgen-Nahkanten-Absorptions-Spektroskopie legt nahe, dass nach 24 Stunden mechanischen Legierens nur 10–14% der ursprünglich zugesetzten 0.3 Gew.% Y_2O_3 in der Stahlmatrix aufgelöst wurden. Die Analyse der Erweiterten Röntgenabsorption Feinstrukturspektroskopie (EXAFS) weist auf eine Koexistenz von Y_2O_3 und $YCrO_3$ in den Legierungen ohne Titan hin. Dies wurde durch EDX Mappings bereits validiert. Die stärker dispergierten yttriumhaltigen Oxide in den kompaktierten ODS Stählen sind ein klarer Hinweis auf die Koexistenz von kristallinen Nanopartikeln und ungeordneten Phasen nach der Hochtemperaturkonsolidierung und –glühung.

Contents

Contents

Contents

List of abbreviations

APT	Atom Probe Tomography
bcc	body-centered cubic
BPR	Ball-to-Powder Ratio
DBTT	Ductile–Brittle Transition Temperature
DEMO	DEMOnstration Power Plant
D-T	Deuterium-Tritium
EBSD	Electron Backscatter Diffraction
EDX	Energy-Dispersive X-ray spectroscopy
EELS	Electron Energy Loss Spectroscopy
ESRF	European Synchrotron Radiation Facility
EXAFS	Extended X-ray Absorption Fine Structure
fcc	face-centered cubic
FT	Fourier Transform
HAADF	High Angle Annular Dark-Field
HAGB	High Angle Grain Boundary
HCLL	Helium-Cooled Lithium-Lead
HCPB	Helium-Cooled Pebble-Bed
hcp	hexagonal close-packed
HEMJ	He-cooled Modular divertor with multiple-Jet cooling
HE	Hot Extrusion

HIP	Hot Isostatic Pressing
ICP-OES	Inductively Coupled Plasma- Optical Emission Spectrometer
ITER	International Thermonuclear Experimental Reactor
KAM	Kernel Average Misorientation
LAGB	Low Angle Grain Boundary
LCF	Low Cycle Fatigue
MA	Mechanical Alloying
ND	Normal Direction
ODF	Orientation Distribution Function
ODS	Oxide Dispersion Strengthen
OM	Optical Microscopy
PF	Pole Figure
RAFM	Reduced Activation Ferritic Martensitic
RD	Rolling Direction
RSD	Relative Standard Deviation
SANS	Small-Angle Neutron Scattering
SEM	Scanning Electron Microscopy
SRS	Synchrotron Radiation Source
SSTT	Small Specimen Test Technology
TBB	Tritium Breeding Blanket
TD	Transverse Direction
TEM	Transmission Electron Microscopy
TMP	Thermo-Mechanical Processing

USE	Upper-Shelf Energy
XAFS	X-ray Absorption Fine Structure
XANES	X-ray Absorption Near-Edge Structure
XRD	X-Ray Diffraction

List of symbols

η	Thermal efficiency
W	Output work
Q	Thermal energy
T	Temperature
σ_y	Yield stress
σ_p	Strengthening from particle dispersion
σ_d	Strengthening by forest dislocations
σ_{gb}	Strengthening from grain refinement
σ_{ss}	Strengthening from solid solution
σ_{lf}	Intrinsic lattice friction
T_m	Absolute melting point
d	Grain diameter or particle diameter
k_y	Material constant
$\Delta\tau_{cut}$	Shear stress required for cutting mechanism
γ	Surface energy or austenite
b	Burgers vector
λ	Inter-particle spacing or average mean free path
$\Delta\tau_o$	Shear stress required for Orowan mechanism
G	Shear modulus
$\Delta\sigma_o$	Yield stress required for Orowan mechanism

List of symbols

M — Taylor factor

ρ_{tot} — Total dislocation density

ρ_f — "free" dislocation density

ρ_b — Dislocation density making up subgrain boundaries

t — Thickness or Time

I_t — Total intensity

I_0 — Intensity of the zero loss peak or the incident intensity of X-ray

F — Relativistic correction factor or Force

E_0 — Energy of fast electrons or Threshold energy

E_m — Mean energy loss

β — Collection semi-angle

Z — Average atomic number

r — Thickness reduction ratio

R_a — Roughness average

R_q — Root mean square roughness

R_z — Average roughness depth

R_{max} — Maximum measured roughness depth

ρ — Density

ω — Mass percentage

A — Area or Atomic mass

l — Sample length

σ_m — Mean stress

σ_r	Stress range
σ_a	Stress amplitude
R	Stress ratio or Bonding distance
I	Transmitted intensity of X-ray
μ	Absorption coefficient
$\chi(E)$	EXAFS fine structure function
k	Wave number of the photo-electron
m	Electron mass
$\hbar$	Plank's constant
F_j	Backscattering amplitude
S_0^2	Amplitude reduction factor
N	Coordination number
σ^2	Mean-square disorder of the neighbor distance
δ_j	Phase shift
E	Young modulus
ν	Poisson's ratio

1 Background

The challenges of meeting rapidly growing demand for energy while reducing reliance on fossil fuels have rekindled a worldwide interest in nuclear power systems including both fission and fusion reactors. Advanced nuclear power plants aim for improved thermal efficiency by extending the operational temperature window.

High operational temperature, high energy neutron irradiation, corrosive environment and required prolonged lifetime impose a complex challenge for materials scientists. It is, therefore, very urgent to develop high-performance materials which can accommodate the specific harsh environment. Reduced activation oxide dispersion strengthened (ODS) ferritic steels are presently considered as promising candidates for nuclear fission reactors, fusion reactors and accelerator driven systems for nuclear waste incineration due to the outstanding microstructural stability, extraordinary high temperature property and irradiation resistance. Besides nuclear application, they also show great promise for various high-tech applications, including CO_2-free production of hydrogen, solar thermal energy conversion or generally high temperature applications in corrosive environments. For purposes of this thesis, the background is confined to innovative nuclear applications. Key points including the basic principle, the service conditions and comparison of existing candidate systems are briefly described.

1.1 Fusion facilities

Energy can be produced by combining light nuclei. Since the early 1950s, four magnetic confinement concepts Tokamak, stellarator, pinch, and mirror were carried out internationally and experienced substantial modifications. For the time being, the International Thermonuclear Experimental Reactor (ITER)—the D-T fueled Tokamak is regarded worldwide as the most viable candidate to demonstrate fusion energy generation [1]. In D-T fueled Tokamak concept, a large amount of energy is released by the most efficient fusion reaction between deuterium (D) and tritium (T) [2].

$$\begin{smallmatrix}2\\1\end{smallmatrix}D + \begin{smallmatrix}3\\1\end{smallmatrix}T \rightarrow \begin{smallmatrix}4\\2\end{smallmatrix}He + \begin{smallmatrix}1\\0\end{smallmatrix}n + 17.6MeV \qquad\qquad (1.1)$$

80% of the energy is carried away by the 14 MeV neutrons and 20% by the 3.6 MeV alpha particles.

The construction of ITER, as shown in Figure 1.1, under a global collaboration aims at demonstrating the feasibility of fusion in burning plasmas on the scale of a power plant and of a number of key technologies. Thirty meters in diameter and nearly as many in height, the ITER Tokamak will house a large number of sub-systems and components. One of the great challenges for fusion power is that D-T reaction requires extremely high temperature of 150,000,000 °C to overcome the mutual repulsion between the positively charged nucleuses. The plasma conditions can be sustained by external energy sources (as in existing fusion experiments) or by self-heating of alpha particles produced by fusion reactions.

Materials performance has been recognized as a critical issue by the fusion community. Especially the tritium breeding blankets (TBB) and the divertor, directly facing the hot plasma, are the most critical and challenging in-vessel components for future fusion energy reactor.

The main function of TBB is to breed tritium, transfer neutron kinetic energy to heat energy and provide shielding of vacuum vessel. In Europe, two advanced breeding blankets are currently developed for DEMO reactor specifications, corresponding to the Helium-Cooled Lithium-Lead (HCLL) blanket with a liquid breeder, and the Helium-Cooled Pebble-Bed (HCPB) blanket with lithiated solid breeder [3].

The structural design of a **HCLL** blanket module from the torus equatorial zone is shown in Figure 1.2. Temperature calculations had shown that a maximum first wall temperature beyond 550 °C occurs within an about 2 mm thin layer on the plasma facing first wall surface. This allows a solution on using ferritic steel as base material for the whole blanket structure with a thin ODS layer of 2–3 mm thickness plated onto the plasma facing surface. The respective data for the thermohydraulic layout is provided with more details by Norajitra et al. [3].

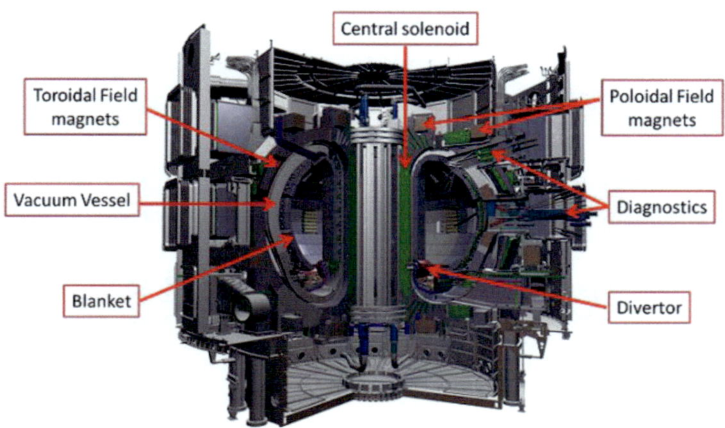

Figure 1.1: International Thermonuclear Experimental Reactor (ITER) [2].

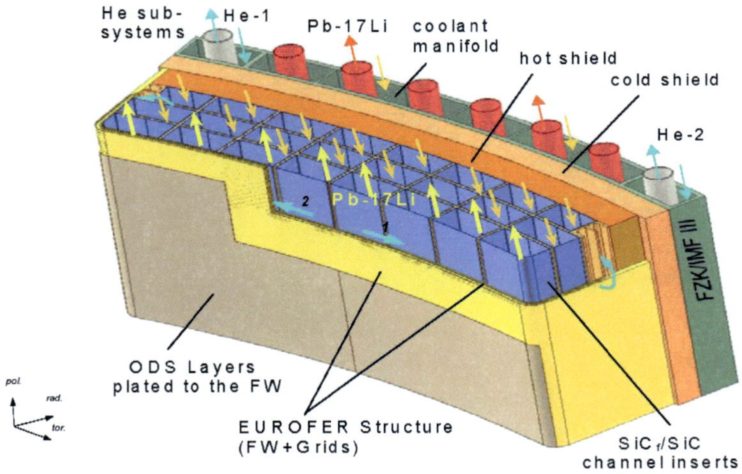

Figure 1.2: Helium-Cooled Lithium-Lead (HCLL) modular blanket design. Its overall radial-toroidal-poloidal dimensions amount to approximately 1 m× 3 m×2 m [3].

The **divertor** plays a positive role to remove the fusion reaction ash (α-particles), unburnt fuel and eroded particles from the reactor. Different divertor types (water-cooled divertor, Helium-cooled divertor and LiPb-cooled divertor) were proposed and the choice of divertor is primarily governed by employing the same coolant with the blanket and operating with a high outlet temperature [4]. Since 2002, two helium-cooled modular divertor designs with jet cooling and with slot array have been systematically investigated in EU. The HEMJ (He-cooled modular divertor with multiple-jet cooling), as shown in Figure 1.3, was chosen as reference concept. It employs small tiles made of tungsten, which are brazed to a thimble made of tungsten alloy W-1% La_2O_3. The W finger units are connected to the main structure of ODS EUROFER steel by means of a copper casting with mechanical interlock. The divertor modules are cooled by helium jets (10MPa, 600 °C) impinging onto the heated inner surface of the thimble [5].

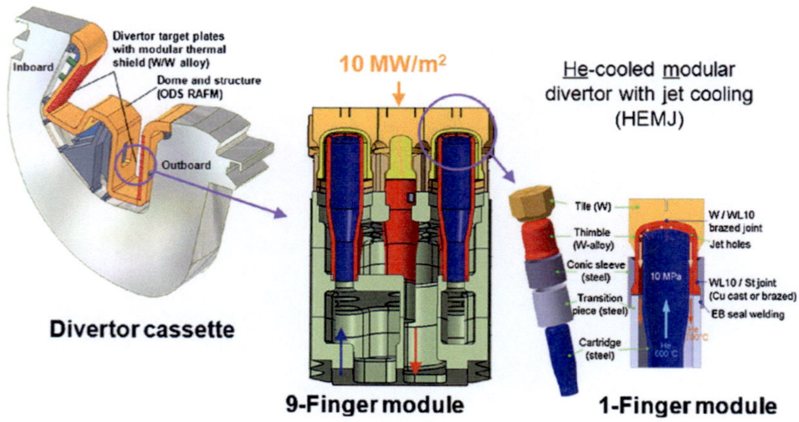

Figure 1.3: The He-cooled modular divertor design HEMJ [5].

Demanding requirements are imposed on materials used in such applications, including low activation, adequate strength and toughness, and high swelling and creep resistance. The development of appropriate structural materials follows at present two distinct lines. The first one is directed towards the construction of ITER which is characterized by a strongly pulsed mode of operation, a very moderate neutron exposure and low operational temperature.

The second developmental line aims for materials which can withstand high neutron fluence and temperature ranges, typical for commercial fusion reactors, and which in addition have the potential for reduced or even low neutron-induced activation.

Only a few material groups can fulfill the requirements and have been pursued in national and international programs. They are advanced ferritic/martensitic steels (ODS steels), vanadium-based alloys, SiC-fiber reinforced ceramic SiC composites and W alloys [6, 7]. Among these candidates, ferritic/martensitic steels are furthest along the development path. There exist a well-developed technology and a broad industrial experience with such alloys in fossil and nuclear energy technology. They show reasonably good thermophysical and mechanical properties, good compatibility with major cooling and breeding materials and a low sensitivity to swelling and helium embrittlement. A major issue, similar to Vanadium based alloy, is the radiation-induced degradation of fracture properties. For further improvement of creep rupture property and higher operational temperature, an advanced variant of reduced activation ferritic martensitic (RAFM) steel with nanoscale oxide dispersions (ODS alloys) has become the focus of international research.

1.2 Fission reactors

Nuclear fission is the splitting of a heavy atom into two or more parts, releasing huge amounts of energy. The release of energy can be controlled and captured for generating electricity. There are currently 441 nuclear power plants in operation around the world, producing 14% of the world's electricity with a total electric capacity of 374.6 GWe [8]. To continue this benefit, 63 plants with an installed capacity of 61 GW are in 15 countries under construction.

To advance nuclear energy for future needs, 13 countries have agreed on a framework for international cooperation in research for a future generation of nuclear energy systems, known as Generation IV (Gen.IV). Six proposed systems refer to Gas-cooled Fast Reactor (GFR), Sodium-cooled Fast Reactor (SFR), Super-Critical Water Reactor (SCWR), Very High Temperature Reactor (VHTR), Lead-cooled Fast Reactor (LFR) and Molten Salt Reactor (MSR). They feature thermal and fast neutron spectra, closed and open fuel cycles and a wide range of reactor size. The goals for Gen.IV systems are sustainability, economics, safety and reliability and physical protection. These requirements result in more severe operation conditions including higher temperature, higher neutron dose and more corrosive environment. A summary of the fuels and materials options (relevant to ODS steels) is provided in Table 1.1 [9].

Table 1.1: The serving conditions, fuel, cladding and structural materials in six Gen. IV reactors.

System	Spectrum T (inlet/)outlet	Fuel	Cladding	Structural Materials	
				In-core	Out-of-core
GFR	Fast: 60dpa 490/850 °C	MC/SiC	Ceramic	Refractory metals, Ceramics, ODS vessel	Ni-based superalloys Or ODS
SFR-metal	Fast, 520 °C	U-Pu-Zr	F/M(HT-9 or ODS)	F/M; 316SS	Ferritic, austenitic
SFR-MOX	Fast, 550 °C	MOX	ODS	F/M; 316SS	Ferritic, austenitic
SCWR-T	Thermal, 550 °C	UO$_2$	F/M steel; Ni-based alloy; ODS	Same as cladding option	F/M
SCWR-F	Fast, 550 °C	MOX, dispersion	F/M steel; Ni-based alloy; ODS	Same as cladding option	F/M
VHTR	Thermal , 1000 °C	TRISO UOC in Graphite compacts; ZrC coating	ZrC coating and surrounding graphite	Graphite PyC, SiC, ZrC Vessel: F/M	Ni-based superalloys Or ODS

Table 1.1 reflects initial suggestions based on experience. Due to the high dimensional stability, outstanding resistance to creep rupture, fatigue crack and the physical and chemical compatibility with the coolant, ODS steels are primary option as fuel claddings for SFR and SCWR and selected as in-core application in GFR.

2 Literature

This chapter reviews the motivation behind the development of ODS steels, the status of understanding and progress and fundamental concepts applicable to their analysis and development.

2.1 Reduced activation ferritic martensitic steels

From the beginning of the attempts to harness fusion power, it was evident that the availability of appropriate materials to operate near the power source is crucial. Early work was aimed at the evolution of the steels that performed well in fission reactors, for instance, the conventional austenitic steels and the low activation variant. Nevertheless, limitations from swelling, inferior thermal properties, helium embrittlement and microstructural instabilities forced researchers to explore other paths [10]. Since the late 1970s, ferritic/martensitic Cr-Mo steels have been considered alternate candidate materials to austenitic stainless steels for first wall and blanket structure applications. FM steels are more swelling resistant and possess higher thermal conductivity and lower thermal expansion than austenitic steels, which provide improved resistance to thermal stresses for a reactor operating in a pulsed mode [11].

Structural materials for fusion application should not produce high levels of long-lived radioactive products and short-lived radioactive products should not produce unacceptable safety consequences (due to high decay heat and/or volatile species that could be released in the event of loss of coolant to the blanket region) [12, 13]. This induces severe limitations on the alloying elements.

Since the mid-1980s, 7–12%Cr reduced activation ferritic martensitic (RAFM) steels have been developed worldwide. The principal approaches adopted in this development are: (a) the replacement of the radiologically undesirable Mo, Nb, and Ni in the existing commercial steels by elements such as W, V, Mn, Ta, and Ti, which have equivalent or similar effects on the constitution and structures, and (b) the removal of the impurities that adversely influence the induced activities when present in low concentrations in the steels. Steels with 7–9% Cr received the greatest attention worldwide. 12%Cr steels often contain the δ-ferrite phase that lowers the fracture toughness. If carbon or manganese is added to the steel to suppress δ-ferrite, either extensive $M_{23}C_6$ precipitates are formed (which tend to reduce the fracture toughness) or the χ phase is formed during irradiation, which has been linked to embrittlement [6, 11, 14].

The composition of several typical RAFM steels is shown in Table 2.1. All the chemical composition in this thesis is expressed by weight percentage (wt. %) without special annotation. Two references F82H and EUROFER have been selected for Japan and Europe respectively. The main compositional differences among these steels include 8–9% Cr and 1–2% W. Higher level of 9% Cr in EUROFER aims at improved corrosion resistance and a minimum shift in ductile-brittle transition temperature (DBTT). The addition of 1% W represents improved tritium breed capacity and a good combination regarding low activation, DBTT, tensile strength, ductility and creep strength. RAFM steel development began with small experimental batches followed by the successful production of the industrial scale batch of F82H and EUROFER 97.

Figure 2.1 clearly shows the progress in RAFM development. The surface gamma dose rate in dependence of the time after irradiation is given. First RAFM steels like OPTIFER reached the remote recycling level at about 100 years while EUROFER and F82H-IEA could be stored as low-level waste after the same period of time after reactor shut-down. EUROFER ref. stands for an alloy composition containing the theoretical values of undesired elements. It seems to be technically feasible to move the activation level into the hand-on-level domain by increased feedstock control and production lines reserved for low activation steel only.

A comprehensive characterization program is being performed on EUROFER 97 including mechanical properties and microstructural stability. EUROFER 97 shows comparable tensile properties with F82H-IEA between RT and 750 °C. Charpy impact tests using ISO-V specimens show a DBTT of -70 °C for EUROFER and slightly higher DBTT of ~40 °C for F82H. Creep-rupture experiments between 450 and 650 °C up to 30000 h show satisfactory results, indicating long-term stabilities and predictability. Aging between 500 and 600 °C up to 10000 h does not significantly influence the tensile and impact properties, confirming high microstructural stability [15].

Table 2.1: Chemical composition of several Japanese and European RAFM steels.

Country	steel	Cr	W	V	Mn	Ta	C	Si	N	B
Japan	F82H	8.0	2.0	0.2	0.50	0.04	0.10	0.10	<0.01	0.003
	F82H achieved	7.5	2.01	0.14	0.17	0.023	0.093	0.16	0.006	
	F82-IEA	8.0	2.0	0.2	0.10		0.10	0.10		
	JLF-1	9.0	2,0	0.2	0.45		0.10	0.08	0.05	
Europe	OPTIFER 1a	9.3	1.0	0.25	0.50		0.10	0.06	0.015	0.006
	OPTIFER II	9.4		0.25	0.50		0.125	0.04	0.015	0.006
	EUROFER	9.0	1.1	0.25	0.40	0.12	0.11	0.05	0.03	0.005
	EUROFER achieved	9.04	1.04	0.19	0.49	0.14	0.109	0.03	0.023	

The irradiation-induced materials degradation is of primary concern for fusion power plant. Irradiation hardening is generally observed by tensile test which is measured as an increase in yield stress and ultimate tensile strength and a decrease in ductility. Irradiation behavior depends on temperature and neutron fluence. For irradiation above 425–450 °C, properties are generally unchanged rather irradiation-enhanced softening occurs, depending on fluence [17, 18]. The effect on toughness is characterized by Charpy impact tests as an increase in DBTT and a decrease in upper-shelf-energy (USE). The shift in DBTT varies inversely with irradiation temperature. Irradiation induced hardening and embrittlement at temperature below 350 °C are the most critical issues for ferritic/martensitic steels [16]. In general, RAFM steel EUROFER exhibits superior irradiation resistance to conventional FM steels (HT9, MANET, and

T91). This offers the possibility of significantly greater flexibility in potential operating temperatures for the reduced-activation steels in fusion structural applications.

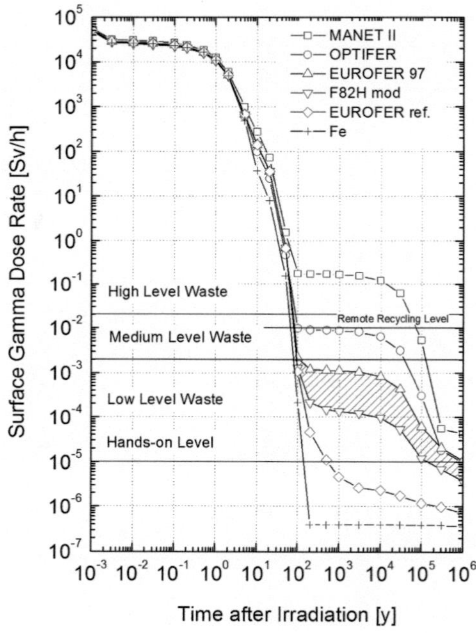

Figure 2.1: Calculated decay of γ-surface dose rate in iron and ferritic/martensitic steels after irradiation (12.5 MWa/m^2) in a first wall DEMO spectrum [16].

Moreover, tungsten inert gas (TIG) welding and electron-beam welding show acceptable results for component manufacturing. On the other hand, hot isostatic pressing (HIP) bonding allows the shaping of components that cannot be made by fusion welding. RAFM steels can be joined defect free by HIP under strict conditions [10]. Friction stir welding (FSW) is relatively new joining technique invented at the Welding Institute, Cambridge, UK in 1991, which is now commercially available. FSW is a solid state welding technology which does not involve melting of the materials and therefore avoids significant changes in microstructure or mechanical properties. The detailed research effort by Chen et al. [19] shows that FSW exhibits a great potential for the joining of ODS materials.

2.2 ODS steels

As noted in the previous section, fruitful progress has been obtained concerning the development of 9% Cr RAFM steels. Their utilization is, however, limited to 550–650 °C due to inferior tensile and creep strength at higher temperature (>700 °C). In order to achieve higher plant operation temperature without losing the merits inherent in RAFM steels, worldwide efforts have been devoted to develop ODS steels. Two key ingredients for ODS steels alloys are (*a*) high, stable dislocation sink strengths and especially large numbers of stable, nanometer-scale precipitates that trap He in fine-scale bubbles to avoid swelling and protect grain boundaries and (*b*) high creep strength, permitting operation at temperatures above the displacement damage regime [20].

Fisher [21] firstly patented a mechanical alloying (MA)/hot extrusion powder processing route to produce 13–25% Cr ODS ferritic steels, which were marketed in the 1980s as International Nickel Company (INCO) MA956 and MA957.

In Japan, research on these alloys by Ukai and co-workers [22], for the advanced fast reactor, has been conducted since 1987. Outstanding approach has been obtained for developing 9% Cr ODS F/M steel and 12% Cr ODS steel [23-25]. Special efforts were made to obtain homogenous microstructure and mechanical properties which are necessary for plate and tubing fabrication. Two kinds of approach had been experimentally explored: α to γ phase transformation for martensitic 9Cr ODS steels especially aiming at radiation resistant alloys and on the other side recrystallization processing for ferritic 12Cr ODS steels aiming at corrosion resistant alloys [23]. ODS alloys designed for FBR with maximum 12% Cr cannot be applied in high-temperature corrosive environment. Therefore, Al added high Cr (14–22%) ODS steels are being prepared as fuel claddings for Generation IV nuclear energy systems such as SCWR [26, 27], SFR [28], and lead bismuth-cooled fast reactor (LFR) [29]. Kimura et al. [30] proposes that the optimal Cr concentration is around 16% by balancing the corrosion resistance and aging embrittlement. An addition of 4% Al is effective to improve corrosion resistance of 16% Cr ODS steel in SCW and Lead-bismuth eutectic (LBE) which is detrimental to high-temperature strength. Creep strength can be improved by adding small amount of zirconium (Zr) or hafnium (Hf).

European programs [31, 32] placed great emphasis on the development of structural materials for fusion reactor on the basis of 9% CrWVTa alloy EUROFER. EUROFER ODS with 0.3% Y_2O_3 allows a substantial increase of the upper operating temperature from 550 °C to 650 °C and extensive mechanical characterization including tensile property, creep property and impact toughness has been conducted on different experimental batches of EUROFER ODS reviewed by Lindau et al. [16]. The biggest progress was made concerning the impact behavior by appropriate thermomechanical treatment. The USE was increased by about 40% and the DBTT shifted from values between +60 and +100 °C for the first generation ODS EUROFER to values between -40 and -80 °C. The operation temperature is, however, limited to around 650 °C for 9% Cr ODS alloys due to α to γ phase transformation. Recently, research on 14% Cr ODS ferritic alloys has become the main trend to extend the operation temperature up to 750–800 °C [32]. Very low transition temperatures and high upper shelf toughness have not yet been achieved for ODS steels, viable approaches are being carried out to avoid the usual high strength–low toughness paradigm.

Research [33, 34] in United States has focused on some commercial and experimental nanostructured ferritic alloys. The mechanical properties of several commercial alloys including MA956, MA957, PM2000 were compared and encouraging creep results were shown for 12% Cr ODS and MA957 [34]. The specific experimental model alloy composition is: 14% Cr, 3% W (a low activation replacement for Mo at mass ratio of ~2:1 to give approximately equal atomic fractions), 0.25% Y_2O_3 and 0.4% Ti which is referred to as 14YWT [33].

2.3 Nanofeatures characterization

Although it has been known for a decade that the extraordinary mechanical properties of ODS steels originate from highly stabilized oxide nanoclusters with a size smaller than 5 nm, the structure of these nanoclusters has not been clarified and remains as one of the most important scientific issues in nuclear materials research [35]. Such atomic-scale characterization requires a broad range of advanced structural characterization tools, including electron microscopy, atom probe tomography, neutron scattering and X-ray scattering [36].

Nanofeatures characterization involves three fundamental questions: what is the chemical composition, what are the crystal and local structures, and what types of defects are present. These basic characteristics are greatly dependent on the alloying composition and the processing history.

There is considerable controversy over the structure of the oxide particles. A wide variety of particle structures has been reported in various alloys. Ukai et al. [37] first recognized the importance of dissolved Ti in refining the Y_2O_3 precipitates, attributing this to $TiO_2+Y_2O_3$ reactions leading to the formation of nanometer-scale Y_2TiO_5 phases. Since then, extensive transmission electron microscopy (TEM) studies have been carried out to characterize ODS steels and to identify Y-enriched precipitates, typically of several nanometers in size [38-41]. Cubic Y_2O_3 particles were found in EUROFER ODS without Ti and showed a strong orientation correlation $[110]_{YO}\,/\!/[111]_{FeCr}$ and $(1\bar{1}\bar{1})_{YO}\,/\!/(1\bar{1}0)_{FeCr}$ [38]. Near stoichiometric complex oxides with a typical size larger than 10 nm, such as Y_2TiO_5 and Y_2TiO_7, were observed in Ti-containing ODS steels by high resolution TEM [42]. A recent study [35] by Cs-corrected TEM provides compelling evidence that the nanoclusters have a defective NaCl structure with a high lattice coherency with the body-centered cubic (bcc) steel matrix. Plenty of point defects as well as strong structural affinity of nanoclusters with the steel matrix seem to be the most important reasons for the unusual stability of the clusters at high temperatures and in intensive neutron irradiation fields.

The analysis of nanoparticles with a radius < 2 nm in various ODS alloy has been equally ambiguous. Miller et al. [43-45] measured a Y: Ti ratio of considerably less than 1 (ranging from 0.25 to 0.5) by atom probe tomography (APT), with an overall metal to oxygen ratio (M:O) of around 1:1, which would correspond to an oxide with TiO type structure. There is considerable variation in the composition of oxide nanoparticles measured by APT, even when the same material is investigated. A common interpretation of the substoichiometeric oxygen content measured by APT is that not all the oxygen is detected during the APT data acquisition [46]. APT results suggest that many ODS particles have a core-shell-type structure [47], which will also affect the apparent composition measurements. The shell structure has also been observed by TEM [41, 48, 49]. The decoration of the Y_2O_3 core with a V and Cr rich shell suggests that the nano-scaled ODS particles are not only welcome trapping and recombination centers for noble gases and irradiation induced defects, respectively, but also effective traps for the long range diffusion and segregation of major alloying elements. It is expected that the trapping of these alloying elements prevents or at least retards significantly the decomposition and subsequent formation of σ or χ phases [50].

Small-angle neutron scattering (SANS) data by Alinger et al. [33] on 14-YWT (Fe-14Cr-2W-0.3Ti) suggested complete dissolution of Y and O during the MA process and precipitation of the oxide nanoparticles during HIP. However, Brocq et al. observed presence of Ti-Y-O in powder directly after MA in an alloy with similar base composition. Brocq et al. also showed that the particles continued to nucleate and grow during subsequent annealing for short time (5 minutes), and suggested that these particles in MA powders formed the basis for the dispersion of particles observed in consolidated materials [51].

The variety of alloy compositions and processing conditions in the previously published studies makes it difficult to determine whether the variability in oxide phase formation is due to differences in composition and processing, or is inherent to the nature and stability of the oxide nanoparticles. Characterizing the microstructure of well controlled alloy systems at various stages during the processing will assist in evaluating the mechanism of formation of the nanoparticles, and will permit better control of the processing conditions. As the manufacturing is a multi-stage process, it is important to understand what influence each stage of the processing has on the final microstructure.

In this study, the microstructure of 13.5Cr2W (0–0.4) Ti0.3Y$_2$O$_3$ was studied using both X-ray absorption fine structure (XAFS) spectroscopy and TEM. The powder was characterized directly after MA, and after consolidation to elucidate the evolution of the oxide nanoparticles and larger grain boundary oxides.

2.4 Strengthening mechanisms in ODS steels

The complex microstructure and the excellent creep resistance in ODS steels indicates a combination of several important strengthening mechanisms including strain hardening (or work hardening), solid solution strengthening, strengthening by grain size reduction and dispersion strengthening. Virtually all strengthening techniques rely on this simple principle: restricting or hindering dislocation motion renders a material harder and stronger. The quantitative individual contribution of these different mechanisms is not clear yet. However, the yield stress σ_y usually consists of several components as seen in Equation 2.1 [52]:

$$\sigma_y = f(\sigma_p + \sigma_d + \sigma_{gb} + \sigma_{ss} + \sigma_{lf})$$

(2.1)

where σ_y refers to yield stress, σ_p stands for strength increment from particle dispersion, σ_d indicates strength enhancement by dislocation density, σ_{gb} is the stress increase due to the reduction of grain size thus the increase of grain boundary, σ_{ss} corresponds to solid solution hardening and σ_{lf} is intrinsic lattice friction which can be neglected at high temperature but not at room temperature as pointed out on the basis of in situ TEM observation [53].

The effectiveness of a strengthening mechanism can be related to the absolute melting point, T_m, of the metal or alloy [54]. It can be inferred from Table 2.2 that, at temperatures greater than 0.5T_m (reactor temperatures at 0.5–0.7T_m), precipitation strengthening and dispersion strengthening are the most viable techniques for imparting high temperature strength to a material. Work hardening effects will be annealed out at relatively low working temperatures. Solid solution strengthening, by tungsten or molybdenum additions, is most likely less effective at high temperatures due to the ability of dislocations to move under non-conservative processes through thermal assistance.

Ultra-fine grain sizes, though they are stable in these alloys at very high temperatures, may ultimately result in high diffusivity paths enabling thermal creep. Classically, the low temperature yield stress σ_y of most polycrystalline materials depends on the grain size d according to the Hall-Petch equation [55]:

$$\sigma_y = \sigma_0 + k_y d^{-1/2} \tag{2.2}$$

Where σ_0 is the friction stress in the absence of grain boundaries, k_y is a positive material constant depending on the resistance of grain boundaries to dislocation movements and d is the grain size. The Hall-Petch relation has been explained by several models such as the pile-up of dislocations ahead of grain boundaries, the grain boundary acting as a source of dislocations, and the influence of grain size on the dislocation density. For normal crystalline materials with $d > 10$ µm, the $d^{-1/2}$ term in Equation 2.2 is negligible. When 10 µm $> d > 100$ nm, the Hall-Petch predication is in reasonable agreement with experimental result. In very small grains with $d < 20$ nm, this mechanism will break down and consequently a threshold value is expected at which a maximum yield stress can be achieved. In other words, the Hall-Petch slope becomes negative below a critical grain size, which is known as the inverse Hall-Petch effect. Furthermore, the transition in the Hall-Petch slope normally occurs for grain sizes between 20 nm and 100 nm [56, 57].

From a basic perspective, precipitation strengthening takes advantage of a specific phase transformation in which an element in solid solution at high temperature will become metastable and exist as a supersaturated solid solution at a lower temperature. In this supersaturated condition, the solute will be rejected and form precipitates. Typically, in iron based materials, this second phase includes carbides, nitrides and several intermetallic compounds. These precipitates impede dislocation motion and thus enhance the alloy strength. The shortcoming of precipitation strengthening is that this strengthening is lost at high temperatures and for long times due to the coarsening or dissolution of the second phase.

The concept of dispersion strengthening involves the introduction of an *insoluble* 2nd phase which persists to the matrix melting point and is the premise behind ODS alloy systems. The introduction of stable oxide particles into the matrix overcomes many drawbacks to using classical precipitation strengthening, but poses some difficulties. Due to significantly different melting points (T_m^{Fe}=1538 °C, T_m^{yttria}=2400 °C), density (ρ^{Fe}=7.87 g/cm^3, ρ^{yttria}=5 g/cm^3) and the thermodynamic insolubility of the dispersoid alloying elements for ODS materials, conventional melt processing techniques are not feasible.

Table 2.2: Summary of strengthening mechanisms [54].

Strengthening mechanism	Effective Temperature	Comments
Work hardening	$\sim 0.3T_m$	Dislocation/dislocation interaction
Grain size refinement	$\sim 0.3T_m$	Grain boundaries/dislocation interaction
Solid solution strengthening	$\sim 0.4T_m$	Lattice strain field/dislocation interaction
Precipitation strengthening	$\sim 0.6T_m$	Metastable 2nd phase particles, impede dislocation motion
Dispersion strengthening	$\sim 0.9T_m$	Insoluble 2nd phase, stabilize grain, subgrain and creep

2.5 Dispersion strengthening theory

High temperature strength and creep resistance of ODS steels can be remarkably improved by uniformly dispersion of extremely fine second-phase particles [34]. The second-phase particles observed in ODS steels fall into two categories. The first class refers to coarse particles larger than 100 nm in size but with low number density. Their chemical composition is substantially different from nanoscale oxide dispersoids and therefore they are referred to as precipitates. Large precipitates have a minor influence on strength and ductility and have seldom been the subject of detailed investigation. In contrast, the second group, nanoscale oxide particles with much higher number density and volume fraction, are considered as the most important contributor for improved high temperature strength, creep resistance and irradiation resistance [50]. They serve as thermodynamically stable obstacles to dislocation motion. Consequently, only the strengthening from nanoscale oxide particles is considered.

Second-phase particles act in two distinct ways to retard the motion of dislocations. They can exist as strong, *impenetrable* particles forcing dislocations to bypass them (Figure 2.2). This type of dislocation–particle interaction is often referred to as Orowan strengthening. On the other hand, they can act as coherent particles through which dislocations can pass, but only at stress levels much above those required to move dislocations through the matrix phase. In both cases, a stress increase is required to move dislocations through a matrix containing a dispersed particle phase. It is likely that in many dispersion strengthened systems, both of these mechanisms operate, with one mechanism typically dominating over the other.

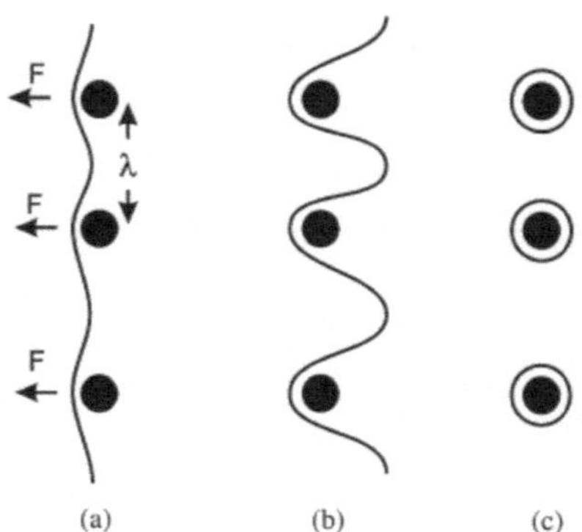

Figure 2.2: The formation of Orowan loops at second-phase particles [58].

The mechanism of particle–dislocation interaction is greatly influenced by the strength of the particle which generally increases with the particle size. When the particle size is smaller than a critical size, the particles are weak and deformable and the cutting mechanism dominates. The stress required for a dislocation to cut through an oxide particle is given by [52]:

$$\Delta \tau_{cut} = \frac{\gamma \pi d}{2 b \lambda} \qquad (2.3)$$

Where γ is the energy of the surface created by dislocation cutting, d is a particle diameter, b is Burgers vector and λ is the inter-particle spacing. A very large value of $\Delta \tau_{cut}$ (about 10 GPa) is roughly estimated by Kubena et al. [52] using γ = 20 J/m^2, d = 2.2 nm and λ = 30–40 nm. The required stress for cutting mechanism is even higher than the theoretical strength for iron which made cutting mechanism energetically unfavorable.

In case of larger incoherent particles, the particles are strong and non-deformable and the particle–dislocation interaction is primarily controlled by Orowan mechanism. The net result of the passage of a matrix dislocation is the generation of extra dislocations in the form of the Orowan loop at the particle. The critical resolved shear stress increase over that for a material without dispersed barriers, $\Delta \tau_o$:

$$\Delta \tau_o = \frac{\alpha G b}{\lambda} \qquad (2.4)$$

Where α is a barrier strength coefficient, G is the shear modulus and b is the Burgers vector. When the barriers are impenetrable α =1 and this is known as the Orowan strengthening. This result shows that the yield stress varies inversely as the spacing between the particles.

The basic Orowan equation has been modified by introducing more refined estimates of the dislocation line tension, by using the planar spacing, and by adding a correction coefficient. These lead to several versions of the equation, of which the most common one is the Orowan–Ashby equation:

$$\Delta\tau_o = \frac{0.13Gb}{\lambda}\ln\frac{r}{b} \qquad (2.5)$$

Note that the particles are assumed to be spherical.

The $\Delta\tau_o$ is the local stress required for a dislocation to move on its slip plane. In order to relate the resolved shear stress ($\Delta\tau_o$) to a polycrystalline material yield stress (σ_y) on a macroscopic scale, an average orientation factor, M, is required:

$$\Delta\sigma_o = M\Delta\tau_o \qquad (2.6)$$

Where M is known as the Taylor factor and 3.06 is the most appropriate value for bcc metals and alloys.

2.6 X-ray absorption fine structure (XAFS)

2.6.1 What is XAFS?

X ray absorption fine structure (XAFS) refers to the details how X rays are absorbed by an atom at energies near and above the core level binding energies of that atom. Specifically, XAFS is the modulation of an atom's X ray absorption probability due to the chemical and physical state of the atom. XAFS is a unique tool for studying, at the atomic and molecular scale, the local structure around selected elements that are contained within a material [82]. XAFS can be applied not only to crystals, but also to materials that possess little or no long range translational order, for instance, amorphous systems, disordered films, liquids, even molecular gases. This versatility allows it to be used in a wide range of disciplines: physics, chemistry, biology, engineering, materials science.

The XAFS spectrum is typically divided into two regimes: X ray absorption near edge structure (XANES) spectroscopy and extended X ray absorption fine structure (EXAFS) spectroscopy. XANES contains the fine structure from the absorption edge to about 50 eV above the edge energy. EXAFS refers to the fine structure from 50 eV to 1000 eV above the edge energy. Though these two have the same physical origin, different approximations, techniques, terminology, and theoretical approaches may be employed in different situations. XANES is strongly sensitive to formal oxidation state and coordination chemistry of the absorbing atom, while the EXAFS is used to determine the distances, coordination number, and species of the neighbors of the absorbing atom.

2.6.2 Physical basics of XAFS

X rays are short wave electromagnetic radiation with energies ranging from 500 eV to 500 keV. There are several types of interaction between X rays and matter, specifically X ray absorption by photoelectric effect, Compton scattering and pair production. Pair production is a high energy phenomenon at very high energies, greater than 1.022 MeV. Therefore it is negligible in the energy range from 3 to 40 keV for XAFS due to the photoelectric effect. The probability for electron scattering is also very small since the X ray wavelength is much larger than the effective electron cross section. Scattering only adds a smooth energy dependent background, which can be easily subtracted afterwards.

Absorption event involves the total absorption of an X-ray photon by an atom as shown in Figure 2.3. As mentioned before, there is a dramatic increase in the probability for the absorption of X-ray with the energy equal to the binding energy of the particular electronic core level. This is referred to as an absorption edge. The excitation of an electron from the most tightly bound shell and the next most tightly bounded shell are termed the K edge and L edge, respectively.

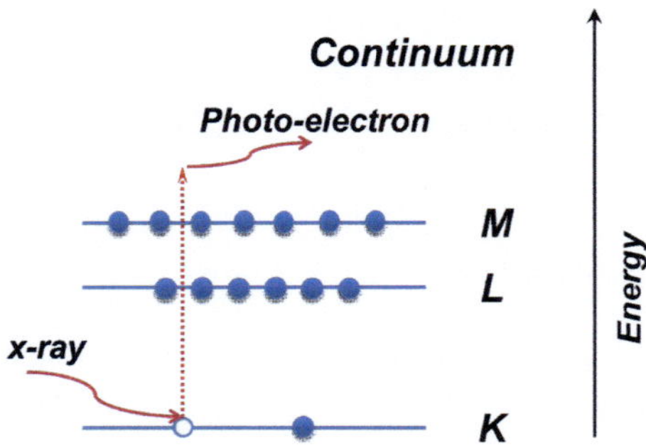

Figure 2.3: A schematic illustration of the photo-electric effect.

After the absorption event, the atom is in an excited state with a core-hole and a photo-electron. The stored energy in the excited state may be released radioactively, by emitting fluorescence radiation. In the process, a core-level electron with higher energy fills the deeper core-hole, ejecting an X-ray of well-defined energy. The fluorescence is characteristic of the atom and can be used to identify the specie of the atoms and to quantify their concentrations. The second process for the de-excitation of the core-hole is the Auger Effect, in which an electron drops form a higher electron level and a second electron is emitted into the continuum. In contrast to the fluorescence process, the Auger electron emission is radiationless, that is, X-ray is emitted by the electron. In both cases, the probability of emission is directly proportional to the absorption probability.

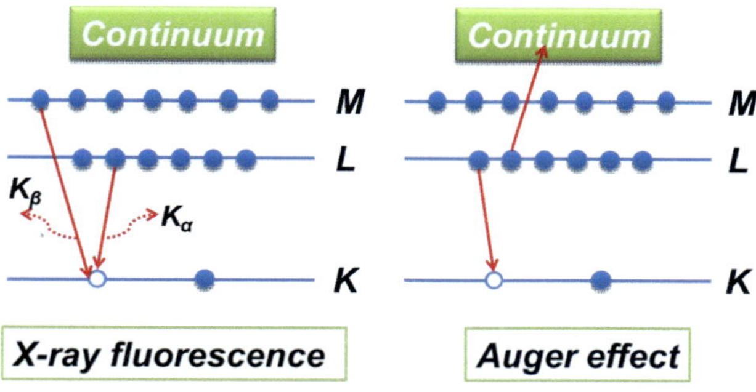

Figure 2.4: A schematic illustration of the X-ray fluorescence and Auger effect.

The intensity of an x-ray beam as it passes through a material of thickness t is given by the absorption coefficient µ. The evolution of µ follows Beer's law:

$$I = I_0 e^{-\mu t} \tag{2.7}$$

Where I_0 is the incident intensity of the X-ray, I is the transmitted intensity and t is the sample thickness.

At most X-ray energies, the absorption coefficient µ is a smooth function of energy, with a value that depends on the sample density ρ, the atomic number Z, atomic mass A, and the X-ray energy E roughly as

$$\mu \approx \frac{\rho Z^4}{AE^3} \tag{2.8}$$

2.6.3 Synchrotron radiation

XAFS requires an intense X-ray beam of finely tunable energy. Therefore, XAFS spectroscopy has developed hand in hand with the growth of synchrotron radiation sources (SRSs). Nearly all modern XAFS experiments are performed at SRSs. Currently there are more than 60 rings of various sizes available worldwide.

SRSs are based on technology originally developed for high energy physics experiments, but subsequently they have been adapted to reliably produce high energy electromagnetic radiation such as X rays with desirable spectral characteristics. Electrons moving close to the speed of light within an evacuated pipe are guided around a closed path of 100–1000 m circumference by vertical magnetic fields. Wherever the trajectory bends, the electrons accelerate (change velocity vector). Accelerating charged particles emit electromagnetic radiation. In this case they profoundly affect the properties of the emitted radiation: the average energy of the X rays and the total radiated power are greatly increased, and the radiation pattern becomes more directional.

In principle, many experimental techniques and sample conditions are available for XAFS, including such possibilities as very fast measurements of in situ chemical process, high spatial resolution and extreme conditions of temperature and pressure. However, the energy ranges, the size and intensity of the beam in practice are dominated by the characteristics of synchrotron sources and experimental station. This often puts practical experimental limits on the XAFS measurements.

2.7 Objective of this work

The study of ODS steels has been long concerned with engineering approaches for the innovative nuclear fission and fusion systems. Recent advances in the science and technology of fusion energy have dramatically improved the prospects of getting the first fusion reactor before mid-century. The harsh requirements in the future fusion reactor have led to a worldwide concentration of developing high-performance ODS materials with an outstanding balance of mechanical, physical and radiological properties.

The significant challenges for the practical development of ODS ferritic steels have been addressed by Odette [20]. In this thesis, the main emphasis will be placed on the following issues:

(1)   Is it possible and how to improve the toughness of ODS steels?

In spite of excellent tensile strength and creep resistance, low transition temperature and high upper-shelf toughness have still not been achieved for

ODS steels. From the literature and the past experience, the strict control of the chemical contamination during mechanical alloying (MA) and the appropriate application of thermomechanical processing (TMP) play a critical role in the toughness improvement of ODS steels.

In this work, the ODS steels will be fabricated by powder metallurgy route including mechanical alloying and hot isostatic pressing. For several specific batches, hot cross rolling will be carried out to obtain homogeneous in-plane properties and enhanced toughness. In order to reduce the concentration of the excess oxygen, the processing of base steel and oxide powders is performed in the Ar-filled glove box. Furthermore, argon is replaced by highly pure hydrogen as the milling atmosphere to avoid the oxygen contamination and the formation of Ar-bubbles in the subsequent compact process. The chemical composition of the ODS steels are monitored after main fabrication steps.

The toughness of ODS steels under various fabrication conditions is characterized by Charpy impact tests.

(2) What is the influence of alloying elements on the microstructure and mechanical properties of ODS steels?

A systematic experimental investigation is carried out to obtain the optimal concentration of the main alloying elements in the ODS steels, such as Cr, W and Ti.

The principal motivation for optimizing Cr content is to get fully ferritic structure for the extended temperature range. However, the Cr content must be controlled to minimize the embrittlement induced by aging or irradiation. 12–14% is selected in this work to obtain fully ferritic ODS steels without introducing brittle Cr-enriched phases.

1–2% W is chosen for a good compromise regarding low activation, DBTT, tensile strength, ductility and creep strength. The tritium breeding ratio is higher for lower W content.

The proper balance of elements such as Ti and excess oxygen is considered as a critical variable for the formation of nanoscale oxide particles. It is reported [20] that high O and Ti ratio enhanced the formation of coarse TiO_2 particles rather than the fine Y-Ti-O features. A slight variation of Ti from 0 to 0.4% is investigated to possess the homogeneously distributed fine Y-Ti-O features and minimize the formation of coarse TiO_2 particles.

The levels of radiologically undesired elements, such as Nb, Mo, Ni, Cu, Al and others, were specified according to the current capabilities of steel making industry in Europe, applying raw materials selection and appropriate clean steel making technologies.

The microstructural characterization of ODS steels involves the joint application of many techniques, including optical microscopy (OM), scanning electron microscopy (SEM), electron backscattered diffraction (EBSD) and transmission electron microscopy (TEM). Vickers hardness, tensile test and Charpy impact test are carried out to evaluate the influence of the alloying elements.

(3)   How does the microstructure evolve prior to and after low cycle fatigue tests?

Low cycle fatigue tests are performed on 13.5Cr1.1W0.3Ti ODS steel after TMP. Careful mechanical polishing and subsequent electro-polishing is applied to improve the surface quality of the fatigue specimen. The fatigue tests are conducted at 550 °C in a strain controlled manner under symmetric tension/compression conditions (R=-1). The microstructure prior to and after low cycle fatigue test, in terms of grain morphology, chemical segregation and dislocation density, are revealed by TEM together with energy dispersive X-ray (EDX) spectrometer and electron energy-loss spectrum (EELS).

(4)   What is the chemical nature of second-phase oxides and how do they evolve during the fabrication process?

This work presents the pioneering attempt to introduce X-ray absorption fine structure (XAFS) spectroscopy to study the evolution of Y-Ti-O features during the fabrication process. The XAFS experiments were performed at European Synchrotron Radiation Facility (ESRF) with an intense X-ray beam of finely tunable energy. Examples of qualitative and quantitative parameters extracted from the dataset are also given and analyzed to show the potential of this technique.

In order to validate the XAFS results, X-ray diffraction (XRD) and TEM are employed to reveal the crystallographic structure and chemical composition of nanoscale oxide particles.

3 Materials and methods

3.1 Raw materials

In this work, ODS steels have been produced on the basis of the gas atomized 12–13.5% Cr base steel powders plus 0.3% Y_2O_3 by powder metallurgy. These 12–13.5% Cr-Ti plus W alloys lie outside the γ-loop, and are therefore classified as ferritic stainless steels, with good corrosion/oxidation resistance deriving from their relatively high Cr content.

3.1.1 Atomized steel powders

Three base ferritic steel powders have nominal composition of 13.5Cr2W, 13.5Cr1.1W0.3Ti and 12Cr2W. These powders were pre-alloyed and produced by inert-gas (argon, Ar) atomization. Generally, high-pressure gas atomization is a close–coupled discrete jet atomization method and is one of the most effective methods of producing powders with high purity and tightly controlled specifications. The chemical composition of these base steel powders is analyzed by the chemical analysis group in our institute.

3.1.2 Dispersion powders

Yttria (Y_2O_3) is selected as the standard dispersion oxide for ODS alloys due to its extremely high temperature stability. Yttrium has an exceptionally high affinity for oxygen, with Gibbs free energy of the oxide formation of 995 kJ mol^{-1} at 1500 K [30]. It is now well recognized that Ti is very effective to refine the dispersion particles and improve the alloy performance. Rather than titanium powder, titanium hydride (TiH_2) is preferred in this work to control the excess oxygen content in ODS alloys.

Y_2O_3 and TiH_2 are provided by Alfa Aesar Johnson Matthey. The basic properties of these powders according to specifications of the powder vendor are summarized in Table 3.1. All these powders are stored in the Ar-filled glove box.

Table 3.1: Basic properties for Y_2O_3 and TiH_2.

Property	Y_2O_3 (Product Nr.: 44048)	TiH_2 (Product Nr.: 89183)
Form	25–50 nm APS Powder	1–3 µm powder
Purity	99.995%	99%
Density	5.01 g/cm^3	3.9 g/cm^3
Melting point	2410 °C	450 °C
Sensitivity		Air & Moisture Sensitive

3.2 Fabrication process

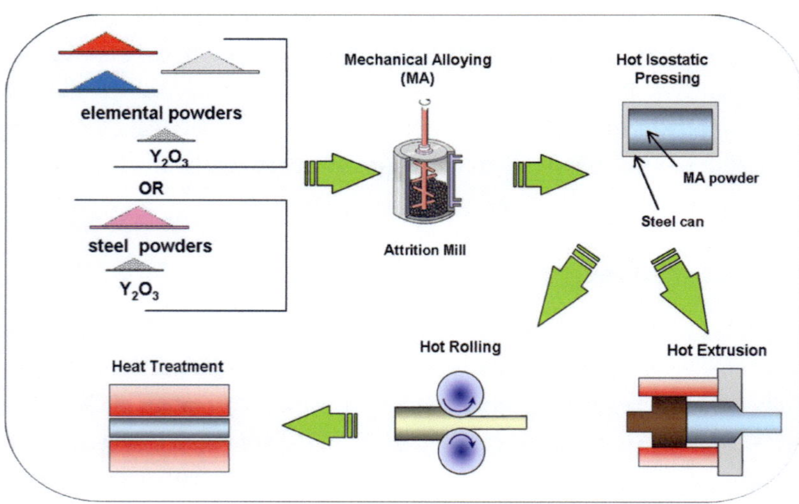

Figure 3.1: The fabrication process of ODS alloys by powder metallurgy.

Conventional ingot metallurgy processing techniques are infeasible for ODS steels due to significantly different melting points (T_m^{Fe}=1538 °C, T_m^{yttria}=2400 °C), buoyancy issues in the melt (ρ^{Fe}=7.87 g/cm^3, ρ^{yttria}=5 g/cm^3) and the thermodynamic insolubility. Although there are several methods to introduce insoluble dispersion, the most effective approach for ODS alloy fabrication is powder metallurgy route consisting of mechanical alloying (MA) of starting powders, followed by consolidation using either hot extrusion (HE) or hot isostatic pressing (HIP), as shown in Figure 3.1. Appropriate thermo-mechanical processing (TMP) is often performed to improve the microstructure and the mechanical properties.

3.2.1 Mechanical alloying

Mechanical alloying is a solid-state processing technique involving cold welding, fracturing, and rewelding of powder particles in a high-energy ball mill. This technique was developed by Benjamin [59, 60] around 1966 to develop oxide dispersion strengthening nickel- and iron-based superalloys intended for applications in the aerospace industry. MA is a simple and versatile technique with important technical advantages. Several generally accepted features of MA are homogeneous dispersion, solid solubility limit extension, alloying of elements with widely different melting points, metastable phase formation and grain size refinement.

(1) Milling equipment

There are different types of mills for conducting MA. These mills differ in their capacity, speed of operation, and their ability to control the operation by varying the temperature of milling and the extent of minimizing the contamination of the powders. Most commonly, the SPEX shaker mills are used for alloy screening purposes. The planetary ball mills or the attritors are used to produce large quantities of the milled powder.

In this work, MA was performed in a ZOZ Simoloyer CM 01 Attritor. As seen in Figure 3.2 a), this Attritor consists of control system, vacuum and gas systems, heating and cooling systems and container & sample units. The high performance control system allows controllable process temperature, controllable milling power, full record of milling history and computer protection of the device. There are many choices for milling atmosphere, for example, in air, in vacuum or in gas (Ar, H_2, or N_2). H_2 is very reactive. The flammability limits of hydrogen in air at 1 atm are 4.0% and 75.0% by volume while explosive limits of hydrogen in air are 18.3% to 59% by volume. In case of H_2 as milling atmosphere, a gas sensor is indispensable to guarantee the safety. Figure 3.2 b) shows the interior structure of a 2-liter stainless steel milling container (X5CrNi18-10, 1.4301) with a rotor and milling balls. The rotational speed of this rotor varies in a quite broad range, from 200 to 1800 rpm. Different kinetic energy can be introduced during MA by selecting various rotational speeds. The milling balls with a diameter of 5 mm are made from Fe-1%C-1.5Cr-0.3Si-1.5Mn steel (100Cr6, 1.3505). During MA, the powerful rotor rotates the impellers, which in turn agitate the steel balls in the container. Collisions between balls, between balls and container wall, and between balls and agitator shaft, and impellers lead to particle size reduction.

Figure 3.2: Milling equipment. a) High-energy ZOZ Simoloyer CM01 attritor; b) Milling container and milling balls.

(2) Main process variables

Milling speed: It is easy to realize that the faster the mill rotates the higher would be the energy input into the powder. But, depending on the design of the mill there are certain limitations to the maximum speed that could be employed. The increased temperature by high milling speed often hinders the powder refining process.

Milling time: The milling time is the most important parameter. Normally the time is chosen in a way to achieve a steady state between the fracturing and cold welding of the powder particles. The time required depends on the type of mill used, the intensity of milling, the ball-to-powder ratio (BPR), and the temperature of milling. The milling time has to be decided for each combination of the above parameters and for the particular powder system. But, it should be realized that the level of contamination increases and some undesirable phases form if the powder is milled for times longer than required. An optimal milling time of 12 h in a planetary ball mill was proposed by Iwata [61] for the fabrication of high-Cr ODS ferritic steels.

Ball-to-powder weight ratio (BPR): The BPR has a significant effect on the time required to achieve a particular phase in the powder being milled. The higher the BPR, the shorter is the time required.

Milling atmosphere: The major effect of the milling atmosphere is on the contamination of the powder. Therefore, the powders are milled in containers that have been either evacuated or filled with an inert gas such as Ar. High-purity Ar is the most common ambient to prevent oxidation and/or contamination of the powder [61-63]. However, nanoscale Ar-filled cavities were found to be preferably trapped at the surface of ODS particles by the EELS with a resolved Ar $L_{3,2}$ edge as well as by the imaging of the single cavity using energy filtered TEM [39]. It was also found that MA in hydrogen yields a significant reduction in oxygen content in the materials, a lower dislocation density, and a strong improvement in the fast fracture properties of the ODS ferritic steels, as measured by Charpy impact tests [62].

(3) Open problems

Powder contamination: A major concern in the processing of metal powders by MA is powder contamination. The small size of the powder particles, availability of large surface area, and formation of new fresh surfaces during milling all contribute to the contamination of the powder. Contamination of metal powders can be traced to (i) chemical purity of the starting powders, (ii) milling atmosphere, (iii) milling equipment (container and grinding medium). The magnitude of contamination depends on the time of milling, the intensity of milling, the atmosphere in which the powder is milled, and difference in the strength/hardness of the powder and the milling medium [64].

Several attempts have been made in recent years to minimize the powder contamination during MA. Firstly, the loading and unloading of the powders into the container is carried out inside atmosphere-controlled glove boxes. Another way of minimizing the contamination from the grinding medium and the container is to use the same or similar material for the container and grinding medium as the powder being milled. In this case, it should be noted that even though there is no contamination, the chemistry of the final powder will be different from the starting powder. Finally, using Ar or H_2 as milling atmosphere is very effective to reduce the excess oxygen content.

In summary, very special precautions should be taken to avoid or minimize the contamination.

3.2.2 Hot isostatic pressing

Hot isostatic pressing is a densification process using heated gas under very high pressure. It involves the simultaneous application of a high pressure and elevated temperature in a specially constructed vessel. Unlike mechanical force, the pressure medium is an inert gas such as argon or nitrogen. It was invented in 1955 for diffusion-bonding application in nuclear industry and then applied to the healing of castings and consolidating powders. HIP process offers several advantages as compared to conventional pressing and sintering process [65]:

Fully dense material can be achieved by removing both macro- and micro- porosity.

Highly complex and near net shape components can be processed.

Homogeneous structure and improved property are obtained.

In this work, HIP was performed by Bodycote. Typical HIP units consist of a pressure vessel, a furnace, a gas system, instrumentation and control and auxiliary systems. Figure 3.3 shows a general layout of HIP chamber. The furnace is fitted on the interior of the pressure chamber for HIP. The wall of the pressure chamber is protected from the heating element by a heat-insulating sheath. Working temperatures are usually between 800 and 1500 °C; in special cases up to 2000 °C. The working pressure resulting from high purity argon is normally from vacuum up to 200 MPa during heating.

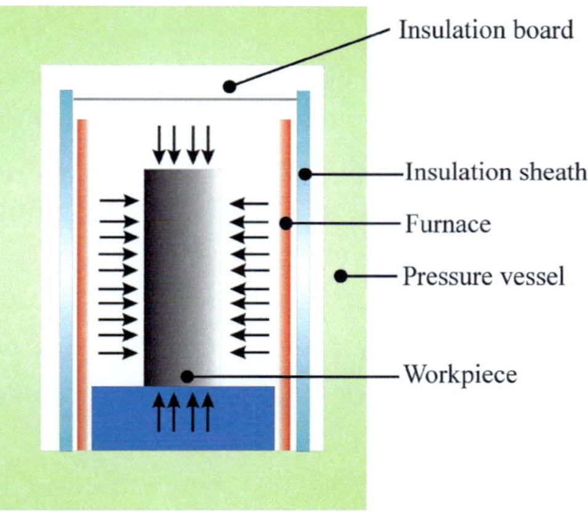

Figure 3.3: Press chamber for hot isostatic pressing [66].

(1) Pre-HIP process (degassing treatment)

Details of these processes were described elsewhere by Eiselt [64]. Briefly, it involves powder filling, degassing and capsule sealing. Firstly, the MA steel powders are placed into a specially designed stainless capsule (1.4301, X5CrNi 18-10) in Ar-filled glove box. The capsule is vibrated to improve the filling density to facilitate the uniform shrinkage during hipping. In order to eliminate the moisture and excess oxygen absorbed by fresh powder surface, these capsules are degassed in a tube furnace at 400 °C for 4 hours until the vacuum of 5×10^{-5} mbar is reached. After degassing, the capsule is vacuum sealed by an electron steel welding tool. Care must be taken in the sealing procedure to prevent leaking of the pressure medium, which may lead to potential contamination and a low density.

(2) HIP process and parameters

HIP involves the application of high temperature and pressure. A cost-effective cycle is employed where the pressure and the temperature are increased simultaneously to reduce the processing time. The holding time starts when the temperature and the pressure reach the target values.

Main parameters in HIP process include temperature, pressing and holding time. The active densification mechanisms are diffusion and plastic deformation. Plastic deformation is a result of the isostatic pressure from the pressurized argon gas within the HIP chamber. For diffusive mechanisms, temperature and time are considered as key variable. The T_m of iron is 1536 °C and typical HIP temperature is usually greater than $0.7T_m$, varying from 950 °C to 1150 °C for steels. SANS study revealed that the number density (N) and volume fraction (f) of nanoclusters in ODS steels are mainly affected by the consolidation temperature. The N and f of nanoclusters decrease and their radius increase with higher consolidation temperatures [33]. High density, pore-free 14YWT nanostructured ferritic steels were processed by hipping mechanically alloyed powders at 1000 °C and 1150 °C. The 14YWT1000 is substantially stronger than 14YWT1150 in terms of micro hardness while the notch fracture toughness is similar for both alloys and insensitive to alloy strength and HIP temperature [67]. Based on the previous study [64], the consolidation temperature of 1150 °C with a pressure of 100 MPa for 2.5 hours is determined in the present work.

3.2.3 Thermo-mechanical processing

Thermo-mechanical processing (TMP) refers to various metal forming processes that involve careful control of thermal and deformation conditions to achieve products with required shape specifications, desired microstructure and properties. Rolling is the most widely used forming process, which provides high production and close control of final product. During rolling, significant thickness reduction can be accomplished by compressive forces through a set of rolls. Two opposing rolls perform two main functions: pull the workpiece into the gap by the friction between the rolls and the workpiece surface and simultaneously squeeze the workpiece. The amount of thickness reduction is termed draft (d) and the reduction (r) is expressed as a fraction of draft over the initial thickness.

$$r = \frac{d}{t_0} = \frac{t_0 - t_f}{t_0}$$ (3.1)

Where t_0 is the initial thickness and t_f is the final thickness.

Hot rolling allows larger amount of deformation and exhibits positive influence on the improvement of impact toughness of ODS RAFM steels by eliminating coarse brittle carbides and homogenizing the microstructure [68]. Multiple passes cross rolling at 1100 °C was applied to a larger batch (1kg) of 13.5Cr1.1W ODS RAF steel. The cross rolling was conducted in three passes, involving two perpendicular rolling directions, with a total thickness reduction (or draft) from 22 mm to 6 mm followed by annealing at 1050 °C for 2 hours. Three passes cross rolling is utilized to symmetrize the crystallographic texture and property anisotropy.

3.2.4 Fabrication parameters

Table 3.2 shows the chemical composition and fabrication parameters of ODS ferritic steels in this work. The MA and HIP parameters were chosen on the basis of the previous study [64]. The present work focuses on the variation of alloying elements (Cr and Ti) and the influence of TMP.

Table 3.2: Chemical compositions and fabrication parameters of ODS ferritic steels.

Capsule Nr.	Chemical composition	MA	HIP	HT/TMP	Remarks
K12	13.5Cr2W	Milling time: 24 hours (1000/4min+700/1min); Milling atmosphere: H_2; Ball-to-powder ratio: 10:1	1150 °C; 100 MPa; 2.5 hours	Annealing: 1050 °C; 2 hours; in vacuum	Ti powder addition
K2	13.5Cr2W0.3Y_2O_3				
K4	13.5Cr2W0.3$Y_2O_3$0.2TiH_2				
K5	13.5Cr2W0.3$Y_2O_3$0.3TiH_2				
K6	13.5Cr2W0.3$Y_2O_3$0.4TiH_2				
K13	13.5Cr1.1W0.3Ti				Ti alloyed
K1	13.5Cr1.1W0.3Ti0.3Y_2O_3				
K14	13.5Cr1.1W0.3Ti			3 pass hot rolling at 1100 °C; annealing	Ti alloyed, 1 kg
K9	13.5Cr1.1W0.3Ti0.3Y_2O_3				
K15	12Cr2W			Annealing, Tempering	
K7	12Cr2W0.3Y_2O_3				
K8	12Cr2W0.3$Y_2O_3$0.3TiH_2				Ti powder addition

Heat treatment parameters for 12Cr ODS steels: annealing (1050 °C/30 min/ vacuum) + air cooling + tempering (750 °C/2 hours/ vacuum)

3.3 Microstructure characterization

3.3.1 Light optical microscopy

Optical microscopy (OM) in materials analysis generally refers to reflected light microscopy. In this method, light is directed vertically through the microscope objective and reflected back through the objective to an eyepiece, view screen, or camera. An optical microscope Reichert MeF3 was used for metallographic study with a magnification up to 1000. This microscope is connected to digital cameras for image capture and subsequent measurement and analysis are performed with a program a4i docu.

Microstructural features of the ODS steels including grain size, material heterogeneity and characteristics of fractured surface can be primarily revealed by OM. The sample preparation procedure is summarized as follows:

Selection of representative samples: For routine inspection, discs were cut from the top, middle and bottom of the hipped capsule to evaluate the materials homogeneity as shown in Figure 3.4.

Choice of surface orientation: For hipped samples, both longitudinal and transverse sections were cut from the top, middle and bottom parts of the capsule. For rolled samples, RD-ND section, RD-TD section and TD-ND section were prepared and investigated to reveal any directional effect.

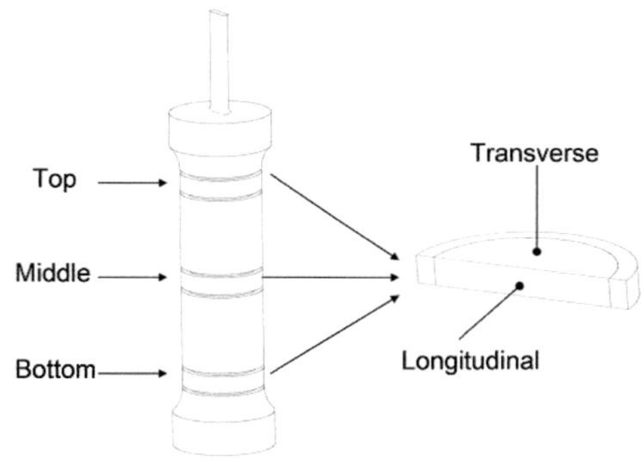

Figure 3.4: Selection of metallographic samples after HIP.

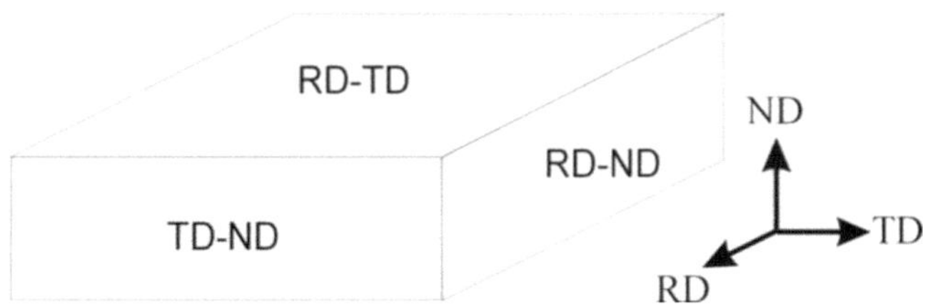

Figure 3.5: Schematic illustration of metallographic samples after rolling. (RD: rolling direction; TD: transverse direction; ND: normal direction).

Proper preparation of sample surface: Sample preparation consists of grinding and then polishing using successively finer abrasives to obtain the desired surface finish at the location of interest before etching. Typically small samples are mounted at room temperature in a mixture of araldite (an epoxy resin), hardener and alumina for better handling. At first stage, the grinding is started with 220 grit SiC grinding paper and finally finished with 1000 grit grinding paper using water as a lubricant to remove the waste. Specimens should be cleaned by detergent and isopropanol after each grinding step to avoid any carryover of abrasive particles to the next step. Proper grinding removes damaged or deformed surface material, while introducing only limited amounts of new deformation. This small amount of damage can be removed by successively diamond polishing and oxide polishing with finer abrasive particles. The grinding and polishing variables are summarized in Table 3.3 and Table 3.4, respectively.

Table 3.3: Grinding parameters for ODS ferritic steels.

No.	Abrasive size [μm]	Time [s]	Force [N]	Rotation speed [min^{-1}]
1	68 (220 grit)	30	90	300
2	46.2 (320 grit)	30	90	300
3	30.2 (500 grit)	30	90	300
4	21.8 (800 grit)	30	90	300
5	18.3 (1000 grit)	30	90	300

Table 3.4: Polishing parameters for ODS ferritic steels.

No.	Abrasive size [μm]	Time [s]	Force [N]	Rotation speed [min^{-1}]
1	9 (diamond)	210	180	150
2	6 (diamond)	120	180	150
3	3 (diamond)	120	180	150
4	<1 μm (OPS)	90	120	150

Selection and control of etchant: 12 Cr ODS samples were etched in a solution of 400 ml ethanol (C_2H_5OH), 50 ml hydrochloric acid (HCl), 50 ml nitric acid (HNO_3) and 6 g trinitrophenol. A solution of 45 ml glycerol, 15 ml HNO_3 and 30 ml HCl was chosen for 13.5 Cr ODS steels. The etching is carried out at room temperature for 3–5 minutes.

3.3.2 Scanning electron microscopy

Scanning electron microscopy (SEM) uses an electron beam rather than visible light for high resolution imaging of surfaces. Sample preparation for microstructural analysis in the SEM was the same as that for OM. In this work, a Philips XL30 field emission SEM is employed. It provides both secondary electron and backscattered electron imaging, along with an Energy Dispersive X-ray analysis (EDX) system. SEM can be operated with an accelerating voltage adjustable between 0.5 and 30 kV. Rapid qualitative chemical information, semi-quantitative composition determinations can be obtained for features or phases as small as about 1 µm by EDX. All the EDX analysis was carried out at an accelerating voltage of 30 kV with a work distance of 10 mm.

3.3.3 Electron backscatter diffraction

Electron backscatter diffraction (EBSD), also known as backscatter Kikuchi diffraction, is an SEM based microstructural-crystallographic technique. It is a powerful tool for crystal orientation mapping, representation of grain size and grain boundary, local texture and discrimination of crystallographically different phases. The basic principles, the sample preparation and the experimental parameters are briefly introduced as follows:

(1) Principles

When an electron beam is incident on a tilted crystalline sample, electron backscatter diffraction patterns are formed on a suitably placed phosphor screen. The diffraction pattern consists of a set of Kikuchi bands which are characteristic of the sample crystal structure and orientation. The center line of each Kikuchi band corresponds to the intersection with the phosphor screen of the diffracting plane responsible for the band. The patterns are transferred from camera to the computer for indexing and determination of crystal orientation, as detailed in Figure 3.6. Briefly, the position of the Kikuchi bands can be found automatically with the Hough transform and used to calculate the crystal orientation of the sample region that formed the pattern. With knowledge of the experimental geometry, the peak locations can be converted to a table of interplanar angles and compared with looking up tables of expected angles for the phases present within the sample. With modern systems, pattern capture and indexing can be performed very rapidly and when combined with precise scanning of the electron beam, a detailed mapping of the sample crystallography (crystal type, crystal orientation, pattern quality etc.) can be obtained. These maps provide rich measurements of the sample microstructure [69].

(2) Sample preparation

The specimen for EBSD was mechanically polished to 18.3 µm (1000 grit) finish. The details of preparation process can be referred to Table 3.3. Micrographs were taken from RD-ND section for the rolled sample.

(3) Experimental

High resolution EBSD system based on field emission gun scanning electron microscopy (FEG-SEM) JEOL JSM-6500F was utilized to resolve texture and grain structure. High resolution EBSD is capable of studying orientation gradient effects at the mesoscopic scale down to a lateral effective resolution of about 20 nm with an angular resolution of about 0.8° under specific conditions [70]. For low pattern quality, however, the angular resolution may be as low as 2°. The measurements were carried out at an accelerating voltage of 15 kV at a specimen inclination angle of 70° towards the camera. Orientation maps were taken at a step size of 200 nm. In the present study, high-angle grain boundaries were defined as interfaces with a misorientation angle of $\theta \geqslant 15°$. Low values of the local misorientation ($2° \leqslant \theta \leqslant 15°$) represent low-angle grain boundary. The experimental setup for EBSD is shown in Figure 3.7.

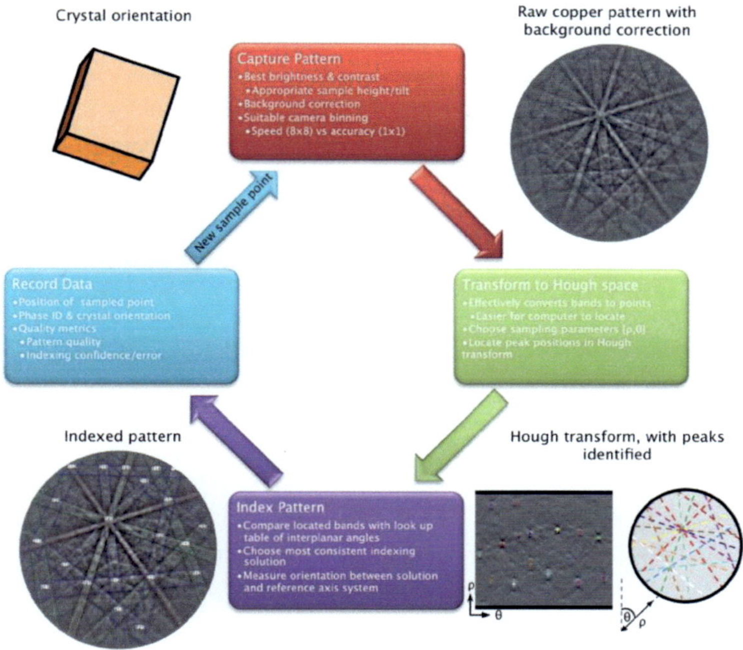

Figure 3.6: Overview of EBSD indexing procedure showing pattern capture through to determination of crystal orientation [69].

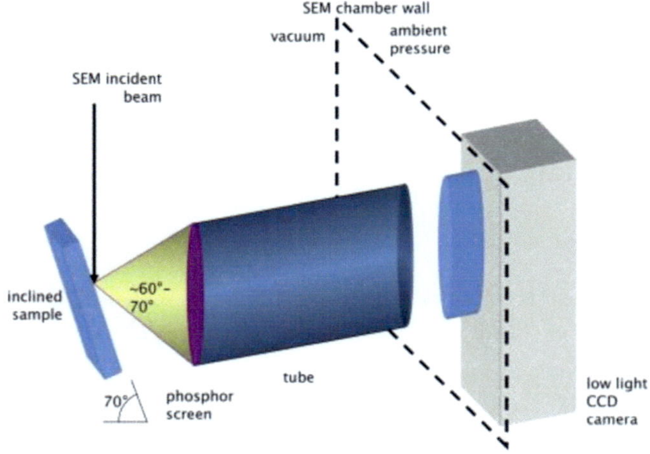

Figure 3.7: Schematic diagram showing the experimental setup for EBSD [69].

3.3.4 Transmission electron microscopy

TEM is of paramount importance to obtain high magnification and high resolution 2D information. The high resolution of TEM enables the structural characterization down to atomic level. TEM offers a tremendous range of signals from which we can obtain images, diffraction pattern and chemical composition characterization from the same small region of the specimen. Therefore TEM is a powerful tool for the investigation of ODS alloys. Only the techniques used in this work are presented.

(1) Principles of X-ray spectrometry and electron energy loss spectrometry

Analytical TEM is capable of measuring the chemical composition by means of X-ray energy-dispersive spectrometer (EDX) and electron energy loss spectrometer (EELS). The EDX is a very sophisticated technique and in principle is sensitive enough to detect all the elements above Li in the periodic table. The major driving force for the development of X-ray microanalysis in the Analytical TEM is the improvement in spatial resolution. This improvement arises since thinner specimen and high accelerating voltage are used in AEM. EELS is the analysis of the energy distribution of electrons that have interacted inelastically with the specimen. The spectrum of electron energies was examined by a magnetic prism spectrometer with resolving power of approximately 1 eV. The EELS technique has come to be an excellent complement to the more widely used X-ray spectrometry, since it is well suited to the detection of light elements which are difficult to analyze with EDX. Despite its simplicity the magnetic prism is operator-intensive and there is not yet the degree of software control to which compared to EDX.

(2) Sample preparation

The TEM specimen, must be thin enough (20–100 nm) to be electron-transparent and representative of the material. Samples with the thickness of 400 µm were cut from the material and mechanically thinned to about 100–120 µm in thickness. 3 mm in diameter discs were punched from the slice and electropolished in a TENUPOL device (TunePol-5, Struers) using H_2SO_4 + CH_3OH as electrolyte. The samples were additionally cleaned for 4 minutes with an Ar ion beam of 2900 V at an angle of 4°. The sample preparation processes is shown in Figure 3.8.

(3) Experimental

TEM investigations were performed using a FEI Tecnai 20F microscope equipped with a Gatan image filter (GIF) for EFTEM and electron energy loss spectroscopy (EELS) measurements, as well as with a high angular dark field (HAADF) detector. The microscope was operated at an accelerating voltage of 200 kV with a field emission gun. TEM images were recorded by means of Gatan charge-coupled device (CCD) camera. EDX and EELS measurements were performed in scanning TEM mode using a high angle annular dark-field (HAADF) detector with a probe size of 1 nm. Such experimental conditions allow reliable particle imaging for particle $\geq$ 2 nm. EDX spectra were recorded using an EDX Si/Li detector with an ultrathin window. And the collection semiangle is 13 mrad.

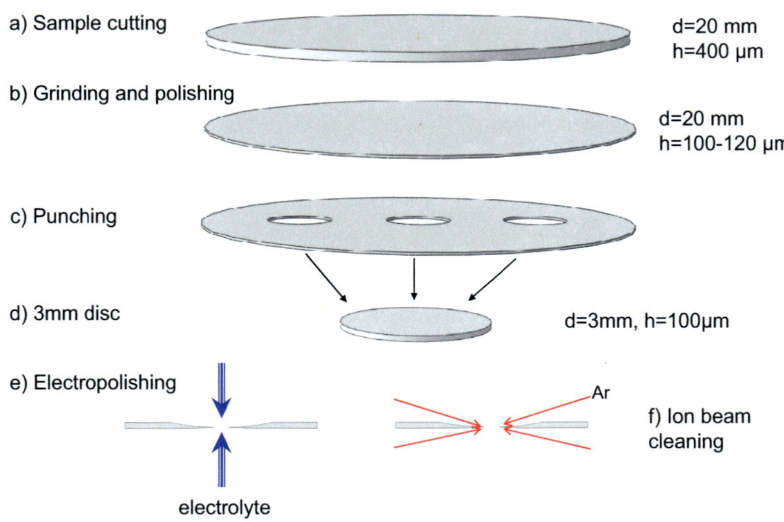

Figure 3.8: TEM sample preparation process.

3.3.5 X-ray diffraction

XRD were performed with Seifert MZ IV diffractometer (Rich. Seifert &Co GmbH). The operation conditions are 40 kV and 30 mA using nickel-filtered Cu Kα1/2 radiations. The diffraction data were collected over a 2θ range of 10° ~ 120°, with a step width of 0.04°. XRD were carried out on the pure Y_2O_3, MA powder and compacted materials.

3.4 Chemical composition analysis

Low activation capacity and mechanical properties of RAF ODS steels can be attributed to the chemical composition during the fabrication process. Precaution should be taken to monitor the pickup of impurities during every step of fabrication. Therefore, 17 elements of the initial steel powder, MA powder, HIP samples and rolled sample are analyzed and compared.

3.4.1 Carbon and Sulfur

Carbon and sulfur contents are determined by a high-frequency infrared carbon and sulfur analyzer (CS600, LECO). Analysis begins by weighing out a sample (1g nominal) into a pre-burned ceramic crucible. Then the combustion chamber is purged with oxygen to drive off residual atmosphere gases. After purging, oxygen flow through the system is restored and heated by an induction furnace. During combustion, all elements of the sample oxidize. Carbon bearing elements and sulfur-bearing elements are reduced, releasing the carbon and sulfur, which immediately binds with the oxygen to form CO or CO_2 and SO_2. Sulfur is measured as SO_2 in first IR cell and CO is converted to CO_2. Carbon is measured as CO_2 in the second IR cell. The results are the average value of two measurements.

3.4.2 Oxygen and Nitrogen

The inert gas fusion oxygen and nitrogen determinator (TC600, LECO) has been used to determine oxygen and nitrogen contents of various materials for many years. The principle of operation is based on fusion of a sample in a high-purity graphite crucible at temperatures up to, or in some cases exceeding, 3000 °C in an inert gas such as helium. The oxygen in the sample will react with the carbon from the crucible to form carbon monoxide (CO); nitrogen is released as molecular nitrogen (N_2). Oxygen is detected either as carbon dioxide (CO_2), CO or both, using infrared detection. Nitrogen is determined using a thermal-conductivity cell.

3.4.3 Metallic elements

Metallic elements in the sample are dissolved into a specific acid and then analyzed by inductively coupled plasma-optical emission spectrometer (ICP-OES) OPTIMA 4300DV, Perkin Elmer. As a matter of routine, representative sample is digested with acids of 7.5 ml mixed acids plus 1.5 ml hydrofluoric acid using a microwave (MARS, CEM GmbH) digestion procedure. The mixed acids consist of 60 vol. % HCl, 20% vol. % HNO_3 and 20 vol. % H_2O. The resulting solution is introduced into the core of ICP, which generates temperature of approximately 8000 °C. Within the plasma, final desolvation, ionization, excitation and characteristic radiative emission for the analytes take place. The resultant emitted radiation is directed through the optics of the spectrometer where it is dispersed via a grating into component wavelengths that are indicative of specific elements present in the plasma. The intensity of the characteristic radiation is measured using a CCD and amplified to yield an intensity that can be converted to an elemental concentration by comparison with calibration standards. The concentration results are the average of three measurements.

3.5 Miscellaneous methods

3.5.1 Particle size determination

Particle size is an important variable to evaluate the influence of MA process. In this work, the low-angle laser light scattering technique (LA950 V2, HORIBA) is selected to measure particle size distribution for both raw steel powder and MA powder. This method is based on diffraction and diffusion phenomenon. A laser beam, when shining through a particle suspension, will be scattered at an angle that is directly related to the particle size. The scattering intensity is also dependent on particle size and diminishes in relation to the cross-sectional area.

Large particles therefore scatter light at narrow angles with high intensity whereas small particles scatter at wider angles but with low intensity. The angle-dependent intensity distribution is recorded by a multi-element detector, and a computer calculates the equivalent particle size distribution.

Before the test, powders are dispersed in 200 ml isopropanol ($CH_3CHOHCH_3$) to separate any adhesive bonds in the particle agglomerates. The uniform distribution of powders in the isopropanol is achieved using ultrasonic homogenizer UW2200 Bandelin for 1 min in pulsed mode (cycle5) with an output power of 40 W. During the measurement, the circulation is always on to avoid powder sedimentation in the sample chamber. Five measurements are conducted for each sample to reduce the statistic error. Samples without ultrasonic homogenizing are also tested and compared.

3.5.2 Dilatometry

Dilatometry is a very sensitive experimental tool to analyze the length (volume) changes and the kinetics of solid state phase transformations during heating and subsequent cooling. A pushrod type dilatometer NETZSCH Dil 402 ED was used to measure the bulk expansion of 12 Cr ODS steel while the sample temperature was measured by a B type thermocouple (Pt30%Rh–Pt6%Rh) mounted in the sample holder [71]. The expansion was measured in the axial direction and the heating/cooling induced length change resulted in the movement of the pushrod. The movement was detected by a linear variable displacement transducer, giving rise to an electrical signal and transmitted to the analog to digital converter. A single crystal of alumina (Al_2O_3) with a length of 25 mm and a diameter of 5 mm was used as reference. The measurement of the thermal expansion was performed in the temperature range from 293 to 1373 K with a heating/cooling rate of 5 K/min in high purity flowing Ar.

3.5.3 Roughness

The surface roughness of fatigue samples was measured with a FRT MicroProf chromatic aberration sensor (CHR 150N). This instrument uses a fast non-contacting high resolution sensor. With a special optical sensor, specimens are illuminated by focused white light. An internal optical device, whose focal length has a strong wavelength dependency, splits the white light into different colors (corresponding to different wavelengths). A miniaturized spectrometer detects the color of the light reflected by the specimen and determines the vertical position on the specimen's surface by means of an internal calibration table. The position is measured with rotary encoders. The device enables measurements with a 10 nm resolution in height and 1–2 mm lateral resolution at a maximum measuring frequency up to 1000 Hz [72]. The recorded roughness parameters are roughly summarized in Table 3.5.

Table 3.5: Main surface roughness parameters.

Term	Definition	Calculation	Application
R_a	Roughness average is the arithmetic average of the absolute values of the roughness profile ordinates	$R_a = \dfrac{1}{l}\int_0^l |Z(x)|\, dx$	The most commonly used roughness parameter; insensitive to distinguish peaks and valleys
R_q	Root mean square (RMS) roughness is the root mean square average of the roughness profile ordinates.	$R_q = \sqrt{\dfrac{1}{l}\int_0^l Z^2(x)\, dx}$	Very similar to R_a, but more sensitive to peaks and valleys
R_z	Average depth is the mean of five roughness depths of five successive sample lengths l	Measured	More sensitive to the single defect since maximum profile heights have been examined.
R_{max}	The maximum roughness depth is the largest among the five roughness depth	Measured	Useful for surfaces where a single defect is not permissible.

3.5.4 Density

The density of the compacted specimens was measured by the water immersion procedure based on the Archimedes principle (ASTM B328-03). For a specific alloy, the theoretical density is expressed in Equation 3.2:

$$\frac{1}{\rho} = \frac{\omega_1\%}{\rho_1} + \frac{\omega_2\%}{\rho_2} + \frac{\omega_3\%}{\rho_3} + \ldots\ldots + \frac{\omega_i\%}{\rho_i} \tag{3.2}$$

Where ρ is the theoretical density of an alloy; ρ_i are the densities of the pure alloying elements and ω_i are the corresponding mass percentages.

3.5.5 Dislocation density

Strain hardening is often utilized commercially to enhance the mechanical properties of steels during fabrication procedures. Strain hardening is largely due to the creation of dislocations during plastic deformation. The hardening can reach saturation once the defect creation and annihilation rates balance. The accurate determination of dislocation density is indispensable in theories of mechanical properties of crystal materials. A wide range of methods for calculating the dislocation density have been developed. It has been shown that the dislocation densities can in principle be evaluated using both X-ray diffraction (XRD) line profile analysis and TEM. The total dislocation density ρ_{tot} is subdivided in two types of dislocations, those which are "free" ρ_f and those which make up sub grain boundaries ρ_b:

$$\rho_{tot} = \rho_f + \rho_b \tag{3.3}$$

It is claimed that ρ_{tot} is only 10% larger than ρ_b which leads to discrepancies when trying to rationalize XRD line profile broadening on the basis of sub grain boundaries. It is well known that sub grain boundaries are associated with short range stress fields while long range stress are required to rationalize XRD line profile broadening.

Quantitative TEM can help to interpret heterogeneous microstructure. However, TEM studies of dislocation densities in ferritic steels are very challenging due to numerous reasons. Firstly, the material is magnetic and consequently the astigmatism of the objective lens must be readjusted for each tilt position. Moreover, high dislocation densities are associated with internal stress which results in diffuse Kikuchi line diffraction patterns. Therefore two beam contrast conditions must be adjusted for each individual micro or nano grain. In some cases dislocation densities can only be evaluated from small regions in a TEM micrograph. For the TEM micrographs evaluated in the present study, a **g** vector of [110] was used. The dislocation density can be obtained as [73]:

$$\rho = \frac{1}{t}\left(\frac{\sum n_v}{\sum L_v} + \frac{\sum n_h}{\sum L_h}\right) \tag{3.4}$$

In this method, a grid consisting of 5 horizontal and 5 vertical lines is superimposed on the central grain of the TEM micrograph. The number n_h and n_v of intersections of dislocations with the horizontal and vertical grid lines are counted. $\sum L_h$ and $\sum L_v$ are the total lengths of the horizontal and vertical test lines. The sample thickness t is measured by electron energy loss spectroscopy (EELS). The EELS spectrum displays an intense zero loss peak due to elastically scattered electrons and low-loss features dominated by the plasmons. The thickness can be calculated from the ratio of the total intensity (I_t) to the intensity of the zero loss peak (I_0) [74]. The expression is

$$t = \lambda \ln\left(\frac{I_t}{I_0}\right) \tag{3.5}$$

The relative intensity ratio is governed by the average mean free path (λ) for energy losses up to 50 eV.

$$\lambda = \frac{106 F E_0}{\left\{E_m \ln\left(\dfrac{2\beta E_0}{E_m}\right)\right\}} \tag{3.6}$$

The average mean free path depends on the relativistic correction factor (F), the energy of the fast electrons (E_0), the collection semi-angle (β) and the mean energy loss (E_m). E_m, for a material of average atomic number Z, is given by

$$E_m = 7.6 Z^{0.36} \tag{3.7}$$

The relativistic factor F is given by

$$F = \frac{\left\{1 + \dfrac{E_0}{1022}\right\}}{\left\{1 + \left(\dfrac{E_0}{511}\right)^2\right\}} \qquad (3.8)$$

The mean free path λ of iron measured with a 200 kV TEM has reported to be 102 nm with the accuracy of ~5%–10% [75].

3.6 X-ray absorption fine structure

3.6.1 Materials and sample preparation

The ODS steel powders were produced by mechanical alloying of pre-alloyed Fe-13.5 Cr-2 W powders with 0.3% Y_2O_3 and 0–0.4% TiH_2 powders. The details of the fabrication process were described in Table 3.2. Briefly, the MA was performed in a ZOZ Attritor, under a highly pure hydrogen atmosphere. Then the milled powders were filled in specially designed stainless steel capsules (18Cr8Ni) and degassed at 400 °C for 4 hours in vacuum. Afterwards the canned powders were consolidated by HIP for 2.5 hours at 1150 °C under a pressure of 100 MPa followed by annealing at 1050 °C for 2 hours. The nominal composition and other experimental details are summarized in Table 3.6. The MA steel powders together with the compacted steels were investigated by XAFS. Commercially available yttria powders and Y metal foil were used as standards.

According to the sample state, there are two types of XAFS samples, powders and compacted samples. For XAFS measurement of the compacted samples, the thickness is the most important feature to optimize the signal to noise ratio. The subsize Charpy impact samples with a thickness of 3 mm were used. An yttrium metal foil with a thickness of 25 mm was measured for energy calibration. The MA powders with very low yttrium concentrations were fitted in a lead holder and then were sealed by adhesive tape which is X-ray transparent. Another reference, the pure Y_2O_3 powder, was mixed with boron nitride (BN) and then ground by the mortar manually. Afterwards the mixed powder was pressed into pellets and measured in transmission mode.

Table 3.6: List of alloy composition, fabrication parameters and measuring mode of XAFS, composition is given in wt. %.

Sample	Composition	Measurement	Remarks
Reference	Y foil	Transmission	purchased
	Y_2O_3		
MA steel	0Ti	Fluorescence	MA:
powders	0.2Ti		24 hours; H_2; Ball-to-powder
Fe-13.5Cr-2W	0.3Ti		ratio=10:1
	0.4Ti		
Compacted	0Ti		MA+HIP+HT:
samples	0.2Ti		1150 °C, 100 MPa, 2.5 hours
Fe-13.5Cr-2W	0.3Ti		(HIP); 1050 °C, 2 hours,
	0.4Ti		vacuum (HT)

3.6.2 Experimental setup

The XAFS experiments were performed on the bending magnet line of sector 26A (BM26A) at the European Synchrotron Radiation Facility (ESRF), Grenoble, France. Figure 3.9 shows the schematic and real XAFS experimental setup. The beam from the bending magnet source passes through a monochromators, which is used to select the X-ray energy. Harmonics are eliminated by a flat mirror installed downstream. Transmission mode and fluorescence mode are two typical modes for XAFS measurements. Transmission mode is the most straightforward: it simply involves measuring the X-ray flux before and after the beam is passed through a uniform sample. The ion chambers are used to record the incident and transmitted X-ray intensities. In fluorescence mode the fluorescence intensity was measured by a multi-element Ge detector. Usually the fluorescence detector is placed at 90° to the incident beam in the horizontal plane, with the sample at an angle (usually 45°) with respect to the beam.

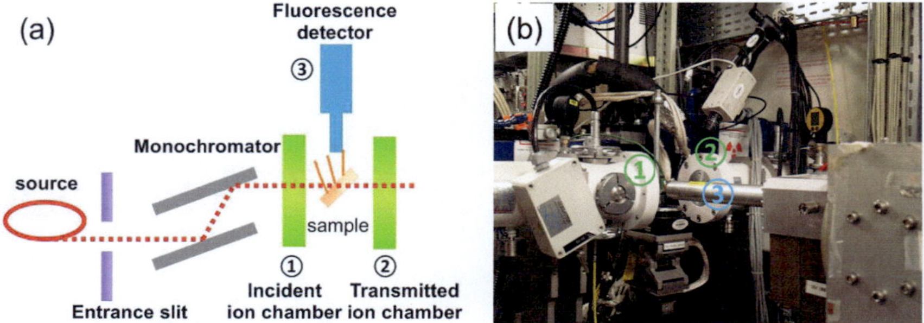

Figure 3.9: XAFS experimental setup (a) a schematic illustration and (b) the real instrument. The incident ion chamber, transmitted ion chamber and fluorescence detector are designated as ①②③, respectively.

As shown in Table 3.6, all data for the MA powders and compacted steel samples were collected in fluorescence mode due to the low concentration of Y. The reference materials, for instance, were measured in transmission mode. All the measurements were performed at Y-edge (17.038 keV) at room temperature. For each alloy sample, 5 to 10 energy scans were collected and averaged to improve the statistics.

3.6.3 Data analysis

The data analysis of XANES and EXAFS are usually separate procedures. EXAFS calculations exhibit better accuracy of quantitative analysis than XANES.

3.6.3.1 XANES analysis

There is a long practice of correlating formal charge state with absorption edge shifts. The K-edge position is the difference between the energy of the 1s level and the final p-state. The final states in the pre-edge are fairly localized. Although small shifts of a few eV may be directly due to other effects, large shifts (10–20 eV) are more often due to chemically induced changes in bond length. In principle, a higher valence of the absorber usually has an absorption edge shifting towards higher energy [76].

It is often useful to fit an unknown spectrum by a linear combination of known spectra. When the sample consists of a mixture of absorbing atoms in inequivalent sites, a weighted average of the $\mu(E)$ corresponding to each site is measured. And the unknown spectrum can be fitted on this basis. If there is no nonlinear distortion, the process of fitting is quite straightforward. Otherwise the distortion must be corrected first. In our work, the relative ratio of metallic Y over total Y is obtained by linear combination fitting of the normalized XANES spectra for MA powders, using metallic Y foil and Y_2O_3 as references.

3.6.3.2 EXAFS analysis

Although the basic physical description of XANES and EXAFS is the same, some important approximations and limits allow us to interpret the extended spectra in a more quantitative way than is currently possibilities for the near-edge spectra.

The standard procedure for EXAFS analysis includes the conversion of the raw data to $\mu(E)$ spectra, background subtraction, normalization, transformation from energy space to wave-vector (k) space, spline function fit, Fourier transforming, filtering, and fitting [76, 77].

(1) Conversion to $\mu(E)$

EXAFS fine structure function $\chi(E)$ is defined as

$$\chi(E) = \frac{\mu(E) - \mu_0(E)}{\Delta\mu_0(E)} \tag{3.9}$$

Where $\mu(E)$ is the measured absorption coefficient, $\mu_0(E)$ is a smooth background function presenting the absorption of an isolated atom, and $\Delta\mu_0(E)$ is the measured jump in the absorption at the threshold energy E_0. The threshold energy E_0 was determined as the energy of the maximum derivative of $\mu(E)$.

(2) Background subtraction

For the background subtraction, a linear function and a two-order polynomial function were used for the fit of the pre-edge and post-edge regions.

(3) Transformation to k-space

EXAFS is best understood in terms of the wave behavior of the photo-electron. It is common to convert the X-ray energy to k, the wave number of the photo-electron. k has dimension of 1/distance and is defined as

$$k = \sqrt{\frac{2m(E - E_0)}{\hbar^2}} \tag{3.10}$$

Where m is the electron mass and $\hbar$ is Plank's constant. Then the primary quantity for EXAFS is $\chi(k)$, the oscillations as a function of photo-electron wave number. To emphasize the oscillations, $\chi(k)$ is often multiplied by a power of k typically k^3. The different frequencies in $\chi(k)$ correspond to different near-neighbor coordination shells according to the basic EXAFS equation.

$$\chi(k) = \sum_j \frac{N_j S_0^2 F_j(k) e^{-2R_j/\lambda(k)} e^{-2k^2 \sigma_j^2}}{kR_j^2} \sin[2kR_j + \delta_j(k)] \tag{3.11}$$

Where the sine function describes the interference pattern. Fj (k) is the backscattering amplitude, which is element specific. S_0^2 is an amplitude reduction factor representing many-body effects, which was experimentally determined by the fit to the model compounds. The exponential term containing the mean free path λ of the photoelectrons accounts for the finite lifetime of the excited state. Using this equation, some physical parameters can be extracted by fitting the XAFS data: 1) R, the distance between the absorber and the neighboring atoms; 2) N, the coordination number of the neighboring atom; 3) σ^2, the mean-square disorder of the neighbor distance or the fluctuation in R_j. To extract these important parameters, an accurate determination of the scattering amplitude F_j (k) and the phase shifts $\delta_j(k)$ is necessary, which were usually calculated by FEFF code using the crystallographic data from model compounds

(4) Fourier transforming and fitting

The Fourier transform of $k^3\chi(k)$ was calculated with a Bessel window function, resulting in a direct correlation with the radial distribution function in real space. The fitting of the first two or three coordination shells was done in real space by using the phase shift and backscattering amplitude extracted from model compounds using FEFF.

3.7 Mechanical behavior

3.7.1 Vickers hardness

Vickers hardness is a measure of materials' resistance to localized plastic deformation and widely applied in materials characterization. In the present work, Vickers hardness is determined by using a HV30 hardness tester from Stiefelmayer Company. A very small diamond pyramidal indenter with an apical angle of 136° is employed to obtain geometrically similar impressions, irrespective of the size and load. The HV number is determined by the ratio of F/A where F is the force (294 N) applied to diamond and A is the contact area of the indenter and material. Hence,

$$HV = \frac{F}{A} = \frac{2F\sin\frac{136°}{2}}{d^2} \approx \frac{1.854F}{d^2} \tag{3.12}$$

Where F is the applied force and d is the average length of the diagonal left by the indenter in millimeters.

All the materials were cut and polished to a mirror surface finish using standard metallographic techniques prior to indentation testing. A flat surface enables the formation of the symmetrical indentation and minimize the testing error. Moreover, 5 measurements were done for each sample to characterize the hardness anisotropy.

3.7.2 Tensile properties

Tensile testing is a fundamental test for ODS steels. The sample is subjected to uniaxial tension until failure. The most important materials parameters, which can be directly measured via a tensile test, are tensile strength, yield strength, total elongation, uniform elongation and reduction in area. Tensile strength is the maximum stress on the engineering stress–strain curve. Generally yield strength indicating the stress for the onset of plastic deformation is very often used for design purposes. The yielding point can be determined as the initial departure from linearity of the stress–strain curve. Consequently, a convention has been established wherein a straight line is constructed parallel to the elastic portion of the stress–strain curve at some specified strain offset, usually 0.002. The stress corresponding to the intersection of this line and the stress–strain curve as it bends over in the plastic region is defined as the yield strength. On the other hand, ductility is a measure of the degree of plastic deformation that has been sustained at fracture. Ductility can be expressed quantitatively as either percent elongation in length or percent reduction in area.

$$\%EL = (\frac{l_f - l_0}{l_0}) \times 100 \qquad (3.13)$$

Where l_f is the fracture length and l_0 is the original gauge length. The elongation at maximum load in a tensile test is denoted as uniform elongation which immediately precedes the onset of necking.

$$\%RA = (\frac{A_0 - A_f}{A_0}) \times 100 \qquad (3.14)$$

Where A_0 is the original cross-sectional area and A_f is the cross-sectional area at the point of fracture

The tensile tests were performed on Small Specimen Test Technology (SSTT) [78] samples with 2 mm in diameter and 7.6 mm in gauge length using Zwick Z030 universal testing machine under vacuum of 8×10^{-7} mbar. The applied strain rate was 2.38×10^{-4} s^{-1} and the elongation was measured by remote-controlled extensometers. The geometry of minimized tensile specimen is shown in Figure 3.10.

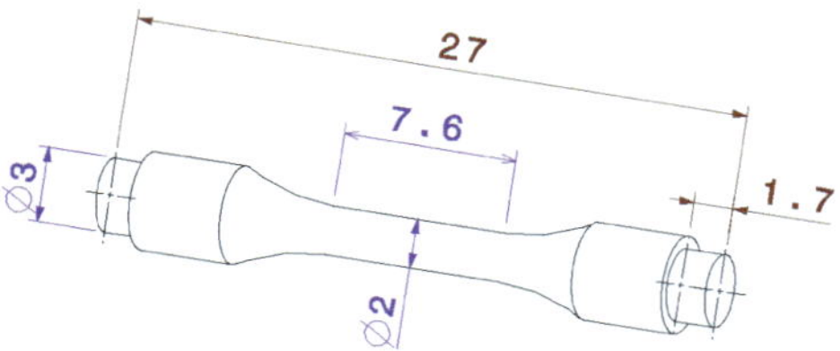

Figure 3.10: The geometry of sub-size tensile specimen.

3.7.3 Impact toughness testing

Tensile tests cannot be extrapolated to fully predict fracture behavior. Therefore, Charpy impact test was established to ascertain the fracture characteristics of ODS steels. The Charpy impact test, also known as the Charpy V-notch test, is a standardized high strain-rate test which determines the amount of energy absorbed during fracture. The absorbed energy is approximately equal to the difference between the potential energies of the hammer before and after impact. The absorbed energy is a measure of notch toughness and acts as a tool to study temperature-dependent ductile-brittle transition. The Charpy test measures the total energy absorbed, which is the sum of the energies dissipated in initiation and propagation of the crack. In view of this problem, the fully instrumented Charpy impact tests were carried out on the sub-sized V-notch KLST specimen (acc. DIN 50115, 3×4×27 mm^3, 1 mm notch depth, 0.1 mm notch root radius and 60° notch angle) in the temperature range from −100 °C to 400 °C. The instrumentation involves the recording of the signal from a load cell on the pendulum in the form of a load–time curve. This curve can provide information about the load at general yield, maximum load and load at fracture. The energy dissipate in impact can be obtained by integration of the load-time curve. However, one must recognize that the result obtained from the Charpy test is only valid in a comparative manner.

The primary aim of Charpy tests is to determine whether or not a material experiences a ductile-to-brittle transition with decreasing temperature. This transition is represented for low strength steels with bcc crystal structure in Figure 3.11. At higher temperatures the impact energy is relatively large, in correlation with a ductile mode of fracture. As the temperature is lowered, the impact energy drops suddenly over a narrow temperature range, below which the energy has a constant but small value. In addition to the ductile-to-brittle transition, the other two general types of impact behavior have been represented by the upper and lower curves of Figure 3.11. It may be noted that low-strength face-centered cubic (fcc) metals and most hexagonal close-packed (hcp) metals retain high impact energy with decreasing temperature. High strength materials, for example ODS steels, are brittle at low temperatures, as reflected by their low impact energy values. The impact energy is relatively insensitive to temperature.

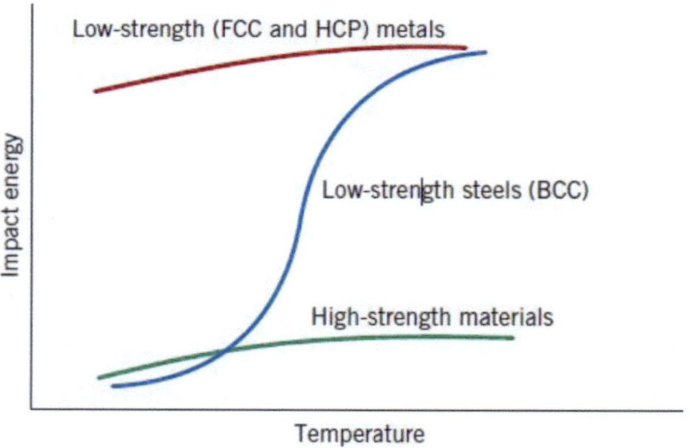

Figure 3.11: Schematic curves for the three general types of impact energy–versus–temperature behavior [79].

3.7.4 Fatigue resistance

3.7.4.1 Basic of Fatigue

In materials science, fatigue is the progressive and localized structural damage under cyclic loading. The nominal maximum stress is less than the tensile stress and even below the yield stress. The pulsed operation and/or maintenance of the future fusion reactor DEMO lead to thermal-fatigue loading of structural materials. Thus the fatigue behavior of ODS steels at elevated temperature is of great significance for the nuclear application.

The total fatigue life is widely used to be characterized as a function of the applied stress range or strain range. The stress- or strain-based methodologies embody the damage evolution, crack nucleation and crack growth stages of fatigue into a single, experimentally characterizable continuum formulation. The stress-life approach is mostly applied in the elastic, unconstrained deformation with low-amplitude cyclic stress for long life. When considerable plastic deformation occurs as a consequence of high stress amplitudes or stress concentrations, the strain-life approach is considered more appropriate for the fatigue tests.

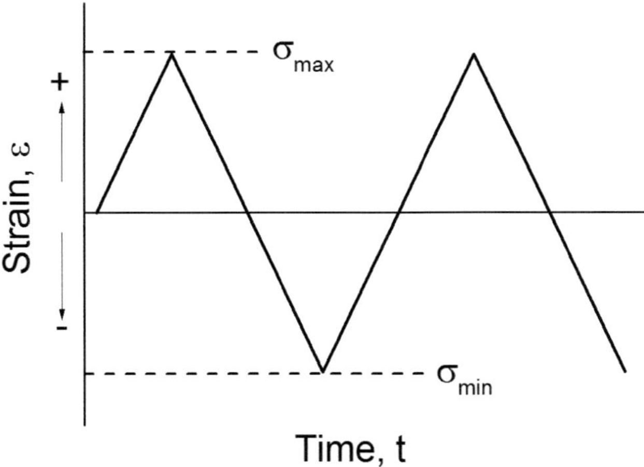

Figure 3.12: Reverse strain cycle for the strain-controlled fatigue test.

The reversed strain cycle was used in this work, as shown in Figure 3.12. Several parameters are used to characterize the fluctuating strain cycle. The strain amplitude alternates about a mean strain ε_m, defined as the average of the maximum and minimum strains in the cycle,

$$\varepsilon_m = \frac{\varepsilon_{max} + \varepsilon_{min}}{2} \tag{3.15}$$

Furthermore, the strain range ε_r is just the difference between ε_{max} and ε_{min}, namely,

$$\varepsilon_r = \varepsilon_{max} - \varepsilon_{min} \tag{3.16}$$

Strain amplitude ε_a is one half of the strain range,

$$\varepsilon_a = \frac{\varepsilon_r}{2} = \frac{\varepsilon_{max} - \varepsilon_{min}}{2} \tag{3.17}$$

The strain ratio R is the ratio of minimum and maximum strain amplitudes,

$$R = \frac{\varepsilon_{min}}{\varepsilon_{max}} \tag{3.18}$$

By convention, tensile strains are positive and compressive strains are negative.

Coffin and Manson proposed a characterization of fatigue life based on the plastic strain amplitude. They noted that when the logarithm of the plastic strain amplitude $\Delta\varepsilon_p/2$ was plotted against the logarithm of the number of load reversals to failure, $2N_f$, a linear relationship resulted for metallic materials,

$$\frac{\Delta\varepsilon_p}{2} = \varepsilon_f'(2N_f)^c \tag{3.19}$$

Where ε_f' is the fatigue ductility coefficient and c is the fatigue ductility exponent [80].

3.7.4.2 Experimental

Miniaturized cylindrical specimen with a gauge length of 7.6 mm at a diameter of 2 mm was proposed in this work. Careful mechanical polishing and subsequent electro-polishing was applied to improve the surface quality of the specimen, thus minimize the lifetime scattering caused by surface defects. The polishing processes consisting of mechanical polishing and electropolishing are summarized in Table 3.7 and Table 3.8. The surface quality of the samples was characterized by SEM as well as surface roughness measurement.

Table 3.7: The mechanical polishing procedures for fatigue specimens.

No.	Abrasive size [μm]	Remarks
1	18.3 (1000 grit)	Grinding paper
2	15	Thread and diamond paste
3	9	Thread and diamond paste
4	6	Thread and diamond paste

Table 3.8: The main parameters for electropolishing process.

Electrolyte	Cathode	Voltage	Time
20 vol. % H_2SO_4+80 vol. % CH_3OH	V4A-Blech	12V	1–2 min

A conventional material testing machine additionally equipped with a load cell, a high performance extensometer and a high vacuum chamber with integrated heating device was used for the low cycle fatigue (LCF) experiments. These specimens were equipped with threaded ends, allowing them to be fixed in a reliable manner. The fatigue tests were conducted at 550 °C in a strain controlled manner under symmetric tension/compression conditions (R=-1). The total strain range was controlled from 0.54% to 0.9%.

4 Results

4.1 Structural analysis

The microstructural evolution of ODS steels are tracked by various microscopes especially electron microscopes during the fabrication process aiming to gain better understanding of processing-structure-property correlation.

4.1.1 Mechanically alloyed powders

4.1.1.1 Powder morphology

The morphology of pure Y_2O_3 powders is shown in Figure 4.1. Y_2O_3 clusters are visible as a mixture of needles and small plates with a size 1–4 µm. This is inconsistent with the particle size 25–50 nm described in 3.1.2. In practice, the individual Y_2O_3 cannot be resolved by SEM because the nanoscale Y_2O_3 exceeds the resolution limit of SEM. Due to the severe agglomeration, the needle-like and rectangular clusters consist of many nanoscale Y_2O_3 particles.

Figure 4.1: Scanning electron micrographs of pure yttria powder.

Moreover, the morphology of base steel powder as well as the MA powders is investigated in SEM, as shown in Figure 4.2. Initially, the base steel powders are found to be spherical with a mean particle size of ~20 µm. The raw steel powders generally have a smooth surface and are free from gas holes. Figure 4.2b shows the particles after 24 hours milling in an attritor with a BPR of 10:1. As mentioned in section 3.2.1, the milling time is the most important parameter for MA and 24 hours is chosen to attain steady-state equilibrium between cold welding and fracturing. After 24 hours milling, the particles evolved from spheres to multiple-layer flakes with irregular shape. A broad range of particle sizes develops with particle size up to 60 µm. Large amounts of cracks are observed on the particle surface which are introduced by heavy deformation during MA. However, even after 24 hours milling, a few of the initial spherical particles can still be found in MA powders.

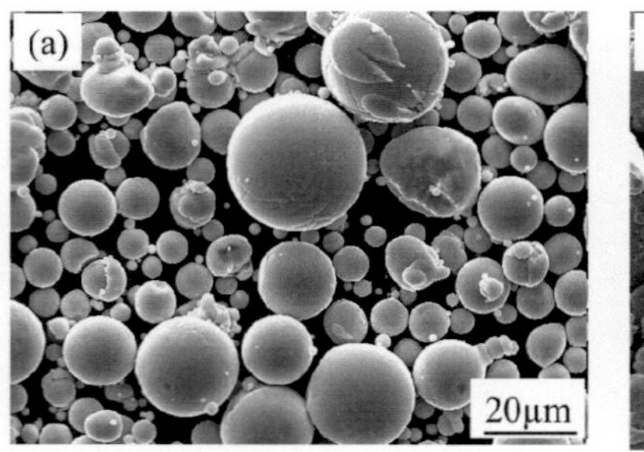

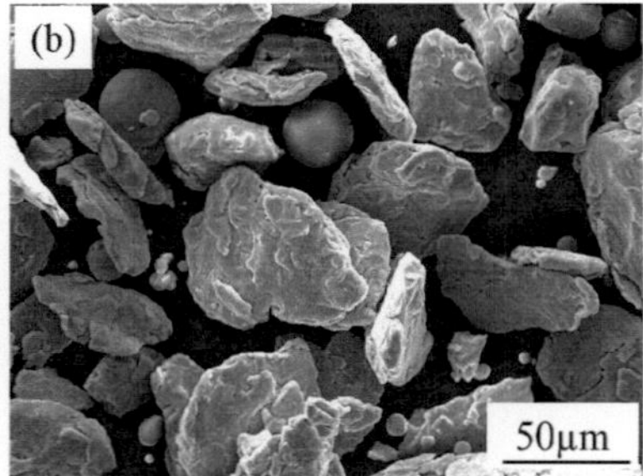

Figure 4.2: Scanning electron micrographs for (a) initial atomized steel powder 13.5Cr2W (K12) and (b) powder 13.5Cr2W0.3Y$_2$O$_3$0.3TiH$_2$ (K5) after 24 hours MA.

4.1.1.2 Particle size distribution

The particle size distribution of the powders was measured by a laser scattering particle size analyzer. Figure 4.3 shows the particle size distribution and the cumulative particle size distribution for two alloy systems before and after 24 hours MA. Each curve is the average result of five measurements. Prior to the test, the powders are dispersed into isopropanol to alleviate the powder agglomeration. The dispersion treatment does not exhibit significant influence on the particle size distribution except for the 13.5Cr1.1W atomized powder. A remarkable shift of mass-median-diameter D50 from 39 µm to 20 µm is observed in Figure 4.3c and Figure 4.3d after the dispersion indicating severe agglomerations in the 13.5Cr1.1W atomized powder. 24 hours MA led to a pronounced particle size increase for both alloys by a factor of 6. This is in good agreement with the SEM observation. The distinct increase in particle size can be explained by the great tendency of cold welding in ductile Fe system during the MA. Another interesting feature is bimodal particle size distribution observed in 13.5Cr1.1W MA powder, two peaks with average diameters of 100 µm and 200 µm. The critical particle size parameters, D10, D50 and D90 of the powders are shown in Table 4.1.

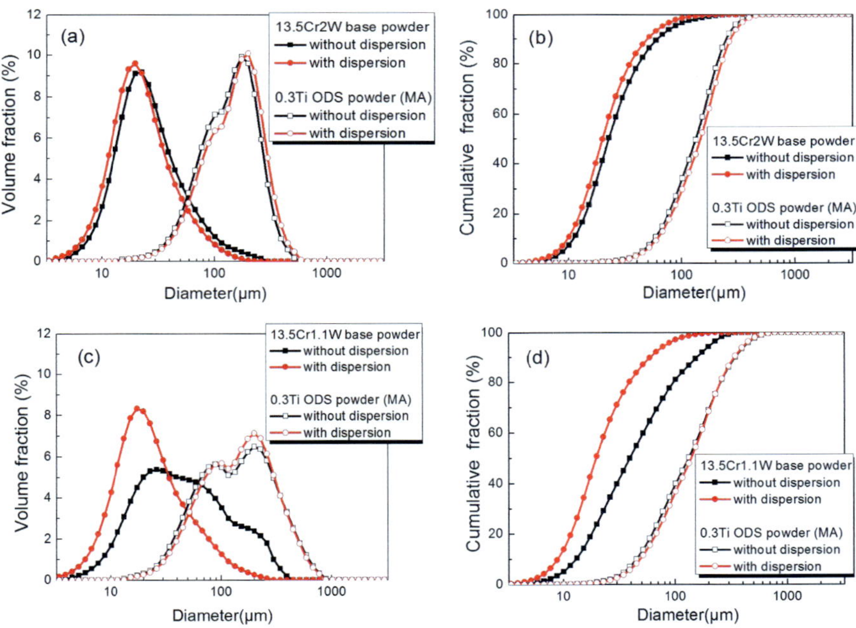

Figure 4.3: The influence of MA on (a) particle size distribution, (b) cumulative size distribution of 13.5Cr2W powders and (c) particle size distribution, (d) cumulative size distribution of 13.5Cr1.1W powders.

Table 4.1: Particle size of the powders before and after the MA with dispersion treatment.

Sample	D10 [µm]	D50 [µm]	D90 [µm]
13.5Cr2W	10	20	50
13.5Cr2W ODS (MA)	61	147	272
13.5Cr1.1W	9	20	60
13.5Cr1.1W ODS (MA)	48	139	338

4.1.1.3 XRD

Y_2O_3 is the most important component for ODS steels and is used as an essential reference for the XAFS data evaluation. The sesquioxide forms, α-Y_2O_3 and β- Y_2O_3, are the only known stable solid oxides in the Y-O phase system. α-Y_2O_3 (C-Y_2O_3, space group Ia3) is thermodynamically stable from room temperature up to 2330 °C, where it transforms into β-Y_2O_3. The crystal structure of the high-temperature form, β-Y_2O_3, is believed to be hexagonal (space group $P\bar{3}m1$) by most researchers. Recently, another high temperature form of Y_2O_3, a cubic form with the fluorite-type structure (space group $F3m$), has been reported and confirmed [81]. The stability of Y_2O_3 is also strongly dependent on the particle size. The m-Y_2O_3 (monoclinic, space group C 2/m) and α-Y_2O_3 may coexist if the particle size decreases to nanoscale owing to the increasing importance of surface effects [82]. Figure 4.4 shows XRD pattern of pure yttria. In comparison with the Joint Committee on Powder Diffraction Standards (JCPDS) data card, the experimental data is in good agreement with the body centered cubic phase (pattern number is 00-041-1105, $Ia3$). The main reflections are perfectly indexed with the exception of two small reflections at 2θ of 27.6° and 32.36°. The inconsistency at such angles probably results from impurities in the yttria powders.

The XRD pattern of MA powder and compacted ODS steel are shown in Figure 4.5. Only three main α-Fe reflections are visible in both samples. For MA powder, the yttria peak absence suggests its incorporation in the steel upon homogeneous distribution in the matrix. The yttria particles incorporated inside the Fe matrix would give a weaker signal because of the limited penetration depth of the X-rays. The diffracted beam originates from a thin surface layer and gives limited information about the material below the layer up to 5 μm. Pronounced lower intensity is observed in MA powder in comparison to the compacted alloy due to a drastic increase of defects introduced by MA process. Generally, the dissolved Y, Ti and O tend to reprecipitate during the consolidation process and the reflection of Y_2O_3 or other complex oxides should be visible in compacted ODS steel. This is probably due to the low concentration of Y_2O_3 addition. XRD provides reliable results for a compound with a concentration larger than 1 wt. % and this insufficiency would be complemented by XAFS.

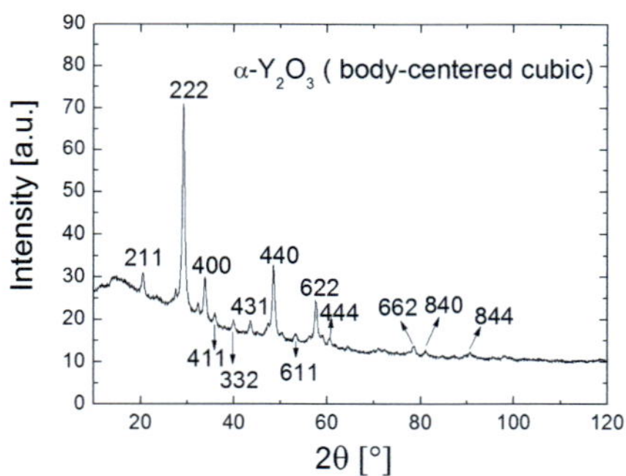

Figure 4.4: XRD pattern of the pure yttria powder.

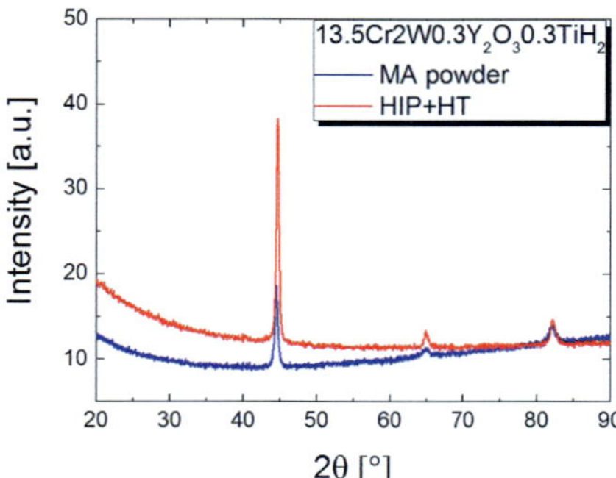

Figure 4.5: XRD pattern of MA powder and compacted ODS steel.

4.1.2 Compacted ODS steels

4.1.2.1 Densities of as-hipped ODS steels

The quality of the compacted ODS steels are firstly evaluated by the density. Low density leads to porous structure and poor mechanical properties. The density measured by the water displacement procedure is listed in Table 4.2. After HIP, nearly fully dense alloys were obtained with relative densities ranging from 99.1% to 99.9%. This finding indicates the compaction by the HIP-cycle was successfully conducted and this was further supported by the absence of porosity observed by OM and SEM.

Table 4.2: The density of as-hipped ODS ferritic steels.

Capsule Nr.	Chemical composition	Theoretical density $[g/cm^3]$	Measured density $[g/cm^3]$	Relative density [%]
K12	13.5Cr2W	7.86	7.79	99.1
K2	$13.5Cr2W0.3Y_2O_3$	7.84	7.80	99.5
K4	$13.5Cr2W0.3Y_2O_30.2TiH_2$	7.83	7.82	99.9
K5	$13.5Cr2W0.3Y_2O_30.3TiH_2$	7.82	7.75	99.1
K6	$13.5Cr2W0.3Y_2O_30.4TiH_2$	7.82	7.78	99.5
K13	$13.5Cr1.1W0.3TiH_2$	7.79	7.73	99.2
K1	$13.5Cr1.1W0.3Y_2O_30.3TiH_2$	7.78	7.77	99.9
K15	12Cr2W	7.87	7.83	99.5
K7	$12Cr2W0.3Y_2O_3$	7.86	7.84	99.7
K8	$12Cr2W0.3Y_2O_30.3TiH_2$	7.84	7.82	99.7

4.1.2.2 The influence of alloying elements

The main difference among these alloys in Table 4.2 is the various concentrations of three significant alloying elements, Cr, W and Ti. The microstructure of the as-hipped ODS steels were investigated by OM, SEM and TEM.

The influence of Ti: 13.5Cr2W ODS steels

The morphology of as-hipped 13.5Cr2W base steel (K12) is shown in Figure 4.6. K12 after HIP is composed of equiaxed grains with a quite broad grain size distribution. The average grain diameter corresponding to Figure 4.6a is 16 µm, which was determined by means of the linear intercept method. The initial atomized powder boundary is indicated by the red dashed circle and a powder particle consists of numerous equiaxed grains. No evident precipitates or porosity are observed in Figure 4.6b.

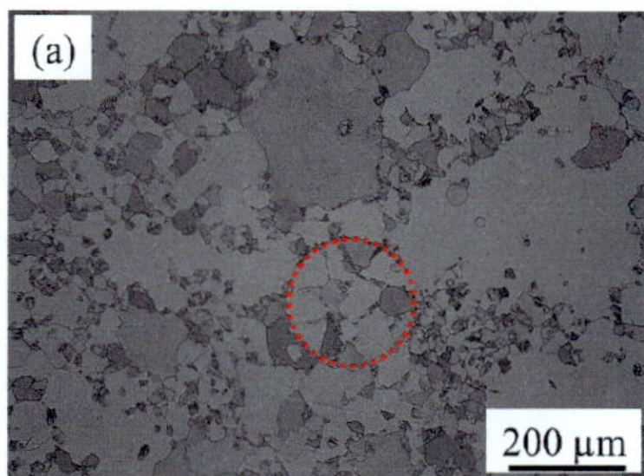
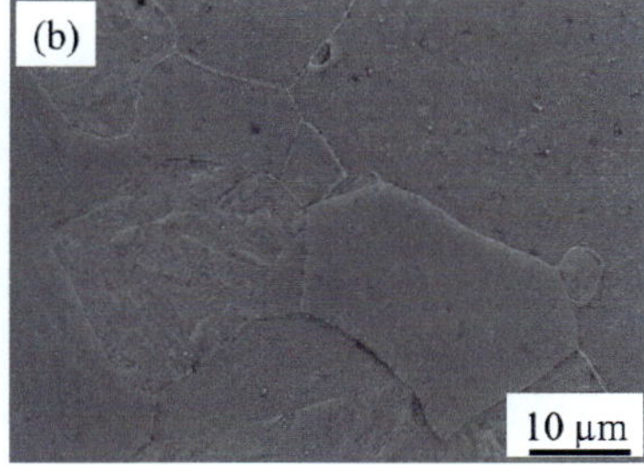

Figure 4.6: Optical micrograph (a) and SEM image (b) illustrating the morphology of as-hipped 13.5Cr2W steel (K12). The red dashed circle points at the initial atomized powder boundary.

Figure 4.7 presents the morphology of the as-hipped 13.5Cr2W ODS steels with different Ti concentrations. All these alloys exhibit heterogeneous microstructure and non-uniform grain size compared to the base steel K12. The grains can be divided into two main groups: micro scale grains with a diameter of 10–20 µm revealed by OM and SEM and possible nanoscale grains indicated by dashed red rectangles. They will be characterized by TEM and shown later. In Ti-free ODS alloy (K2) (Figure 4.7a), there is a distinct boundary between these two grain categories. And the middle part is dominated by the nanoscale grains and the rest mainly consists of micro scale grains. In contrast, grains in Ti-containing alloys (Figure 4.7c, e, g) situate in a more mixed manner and the coarse region is surrounded by the nanoscale grains. Due to this complexity, the average grain size cannot be accurately calculated by the linear intercept method. SEM micrographs (Figure 4.7b, d, f, h) show that a large number of elongated white precipitates align along the grain boundaries. These elongated precipitates in K2 had a typical length and thickness of 1–4 µm and 0.2–0.5 µm, respectively. The estimations show 30–40% of the grain boundaries in K2 are covered with these precipitates. The coverage degree of the grain boundaries was calculated as the ratio of the covered length to the whole length. In addition, an evident decrease in length-to-diameter ratio from 8:1 to 4:1 was found when the titanium content increased from 0% to 0.4%. Despite a homogenous distribution in K2, these long-shaped precipitates distributed very unevenly in 13.5Cr2W0.4Ti ODS steel (K6), some regions showed a high density while others free of them, which make the quantification very difficult.

TEM was employed to further characterize the grain structure of the as-hipped ODS steels. Figure 4.8 shows that K2 exhibits equiaxed grains with a size of 1–8 µm, a low dislocation density and spherical precipitates with a dark contrast. Although nanoscale grains are absent in Figure 4.8, bimodal grain size distribution for K2 cannot be ruled out due to the limited area of the analysis and the microstructural heterogeneity.

In contrast to K2, an evident bimodal grain size distribution of coarse grains, a few micrometers in diameter, and smaller grains, about 200–500 nm in diameter was detected in 0.4Ti ODS alloy (K6) (Figure 4.9). Similar phenomenon is also observed in other Ti-containing alloys K4 and K5, which are not shown here. This bimodal grain structure was quite stable when it was formed and it is very difficult to be eliminated. The underlying reasons are not fully understood yet. Odette et al. [20] suggested that bimodal grain size distribution resulted from the heterogeneous distribution of nanoscale oxide particles. The dislocations recover and coarse grains form in regions with lower particle density, whereas submicron grains with high dislocation densities remains in regions with plenty of oxide particles. This assumption is contradictory to the analytical results. The TEM results by Eiselt [64] clearly reveals a higher particle density in microscaled grains than that in nanograins. The nanograins boundaries are favorable sites for the formation of coarse ODS particles with a diameter greater than 10 nm.

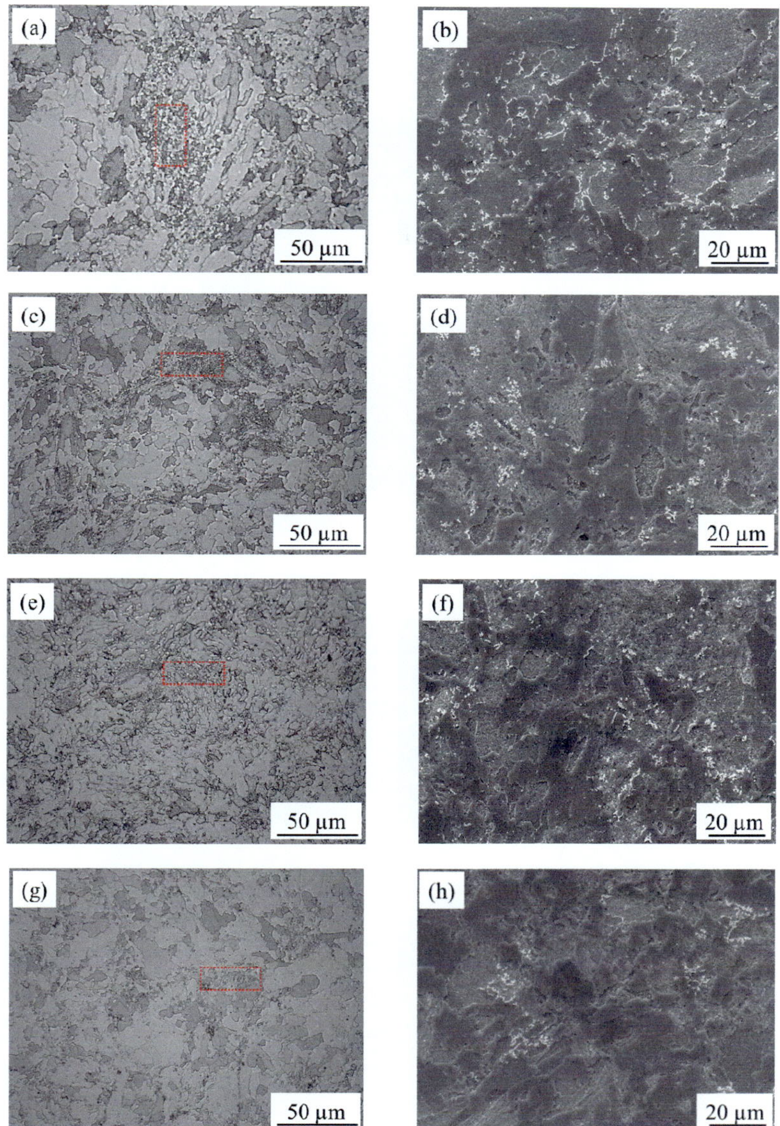

Figure 4.7: Optical micrographs (a) (c) (e) (g) and SEM micrographs (b) (d) (f) (h) for the hipped 13.5Cr2W ODS steels with different Ti concentrations. (a) (b) K2: 0% Ti; (c) (d) K4: 0.2% Ti; (e) (f) K5: 0.3% Ti; (g) (h) K6: 0.4% Ti, respectively. The red dashed rectangles indicate nano grain regions in the compacted ODS steels.

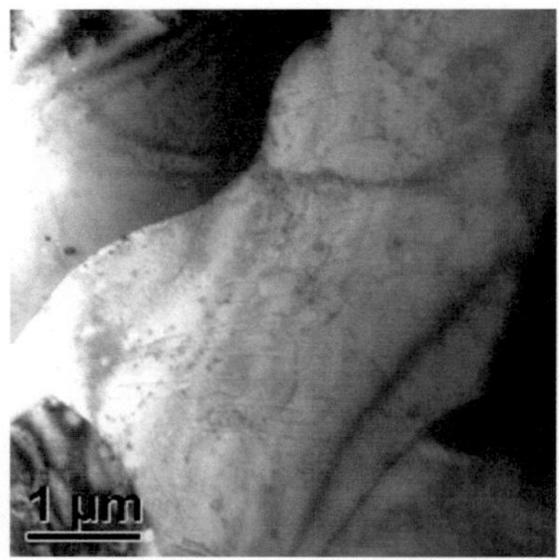

Figure 4.8: Bright field TEM images for the hipped ODS steel K2 (0 Ti).

Figure 4.9: Bright field TEM images for the hipped ODS steel K6 (0.4% Ti). (a) micro grain region, (b) nano grain region.

The influence of W: 13.5Cr1.1W ODS steels

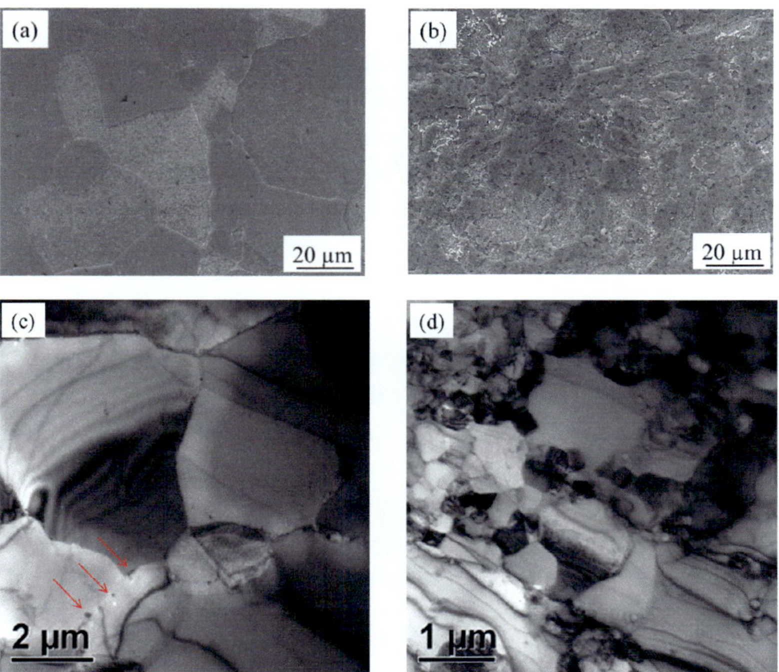

Figure 4.10: SEM micrographs (a) (b) and bright field TEM images (c) (d) for 13.5Cr1.1W base alloy K13 (a) (c) and ODS alloy K1 (b) (d). The red arrows point at the coarse oxides at the grain boundaries.

Figure 4.10 presents the morphology of the as-hipped 13.5Cr1.1W alloys. 13.5Cr1.1W0.3Ti base steel (K13) consists of equiaxed grains with the average diameter of 21 µm indicating the variation in W does not alter the grain morphology. Numerous spherical precipitates indicated by red arrows correspond to Ti oxides, which will be further discussed in section 4.2.2. Contrary to K13, the main features of the ODS steel K1 include uneven distributed elongated Cr carbides, bimodal grain size distribution and high dislocation density within the nanograins.

The influence of Cr: 12Cr2W ODS steels

It is believed that higher Cr concentration leads to pronounced improvement in corrosion resistance. The aging resistance at high temperature and the irradiation resistance under high energy neutrons, on the other hand, degrade significantly due to the re-precipitation of Cr-enriched brittle phases. In principle, 12% Cr is the minimum to obtain pure ferritic steels. However, the alloying elements play a substantial role on expanding or contracting the α-field. For instance, carbon and nitrogen are regarded as the most important γ-stabilizers which encourage the formation of austenite over wider compositional limits. Conversely, silicon, aluminum, beryllium together with the strong carbide-forming elements (titanium, vanadium and chromium) are called α-stabilizers. Due to the pickup of carbon and nitrogen during mechanical alloying, the γ-field expands and ferritic/martensitic microstructure arises in 12Cr ODS steels. Figure 4.11 presents the microstructure of 12Cr2W base alloy (K15). Equiaxed ferritic grains with a diameter up to ~ 40 µm are clearly visible in Figure 4.11. Other regions with significantly darker contrast cannot be resolved by OM. Figure 4.11b reveals martensitic laths with high dislocation density by TEM.

Similar ferritic/martensitic microstructure are found in two 12Cr2W ODS steel K7 and K8. As shown in Figure 4.12, K7 has a smaller grain size (for ferritic grains) and greater martensite portion (approximately estimated 50%). For both materials, the lath boundaries indicated by red dashed lines are almost straight due to the preferred habit plane and the strict crystallographic relationship between the ferrite and the parent austenite. The parallel martensite laths are grouped into martensite units, the units themselves into larger packets.

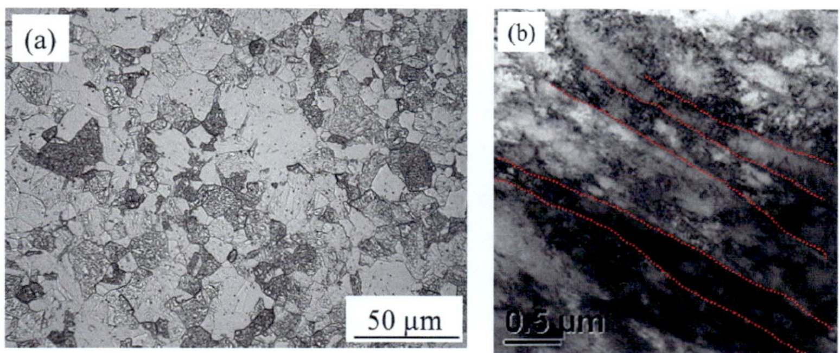

Figure 4.11: OM micrograph (a) and bright field TEM image (b) for 12Cr2W base alloy K15. The red dashed lines indicate the martensitic lath boundaries.

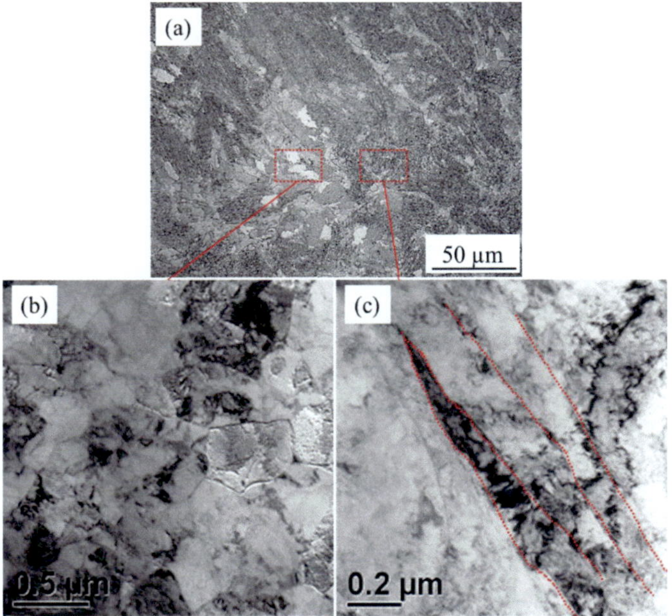

Figure 4.12: OM micrograph (a) and bright field TEM images (b) (c) for 12Cr2W ODS steel K7. The red dashed lines refer to the martensitic lath boundaries.

4.1.2.3 The influence of heat treatment

As pointed out in 4.1.2.2, characteristic features of the as-hipped 13.5Cr ODS steels are bimodal grain size distribution and pronounced elongated Cr carbides, whereas ferritic/martensitic structure is commonly observed in 12Cr ODS steels. Therefore annealing treatment at 1050 °C for 2 hours in vacuum was performed for 13.5Cr ODS steels to homogenize the structure, dissolve the elongated Cr carbide and improve the ductility. Quenching (1050 °C/30 min/ vacuum) plus tempering (750 °C/2 hours/ vacuum) was employed for 12Cr ODS steels due to the presence of martensite laths. The exact quenching and tempering conditions are selected on the basis of the Dilatometry plots of 12 Cr ODS steels.

With respect to the microstructure of the 13Cr ODS steels, W does not play an evident role. Consequently, only the representative micrographs for 13.5Cr2W ODS steels after annealing treatment are shown. Figure 4.13 reveals that annealing treatment is very effective to remove the elongated carbides formed during HIP. The absence of these coarse carbides during annealing indicate they dissolved into the matrix at 1050 °C and there was insufficient time for re-precipitation due to the faster cooling rate in air compared to the slow cooling inside the HIP device. In addition, a small density of spherical precipitates is clearly visible as darker dots in HADDF image with a diameter up to approximately 400 nm and a majority of them are identified as Cr oxides by EDX and EELS. Another characteristic feature is some micro pores. The small residual porosity in compacted ODS alloys is a common problem and accounts mainly for mechanical alloying process. The inert gas argon used during MA accounts for the formation of Ar-filled bubbles in ODS F/M steels. In this work, the replacement of Ar by H_2 during MA enables efficient reduction in oxygen content and stronger diffusion ability from the bulk material to the surface and thus leads to practically fully compacted materials confirmed by high density shown in 4.1.2.1. These micro pores may origin from decohesion of these spherical Cr oxides during the etching process according to the size, shape and distribution of these pores.

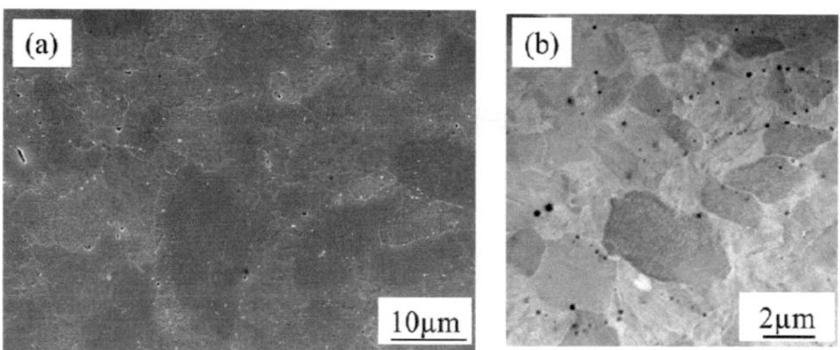

Figure 4.13: SEM micrograph (a) and HAADF TEM images (b) for 13.5Cr2W ODS steel K2.

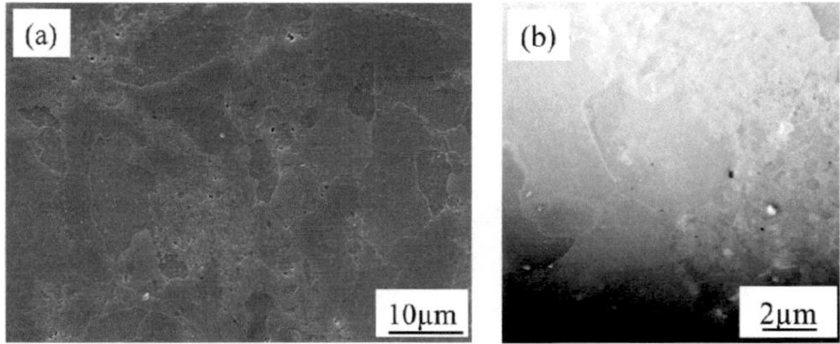

Figure 4.14: SEM micrograph (a) and HAADF TEM images (b) for 13.5Cr2W0.2Ti ODS steel K4.

Figure 4.14 shows the annealed microstructure of 13.5Cr2W0.2Ti ODS steel K4. Cr carbides are absent in comparison to Figure 4.7 (c) and (d) while the bimodal grain size distribution remains unchanged after annealing. This phenomenon is typically observed for all Ti-containing 13.5Cr ODS steels (K4, K5, K6) after annealing. The annealing treatment cannot alter the heterogeneous distribution of nanoscale particles. Therefore, bimodal grain size distribution is very difficult to be eliminated by heat treatment.

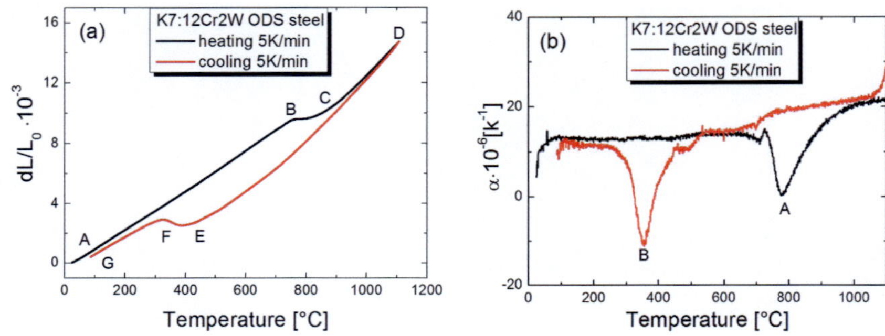

Figure 4.15: (a) Measured length change of 12Cr2W ODS steel during continuous heating and continuous cooling (5 K/min) and (b) Linear coefficient of thermal expansion α as a function of temperature. Initial length of the specimen is 25 mm.

Dilatometry is used to determine the critical temperatures of the phase transformation for 12Cr ODS steels. Figure 4.15 shows the length change and the linear coefficient of thermal expansion α as a function of test temperature for K7. In Figure 4.15a, segment AB corresponds to the normal thermal expansion during continuous heating in the absence of a phase transformation. The plateau of BC represents the $\alpha \rightarrow \gamma$ transformation, during which a length contraction occurs due to the formation of austenite. CD and DE stand for the normal expansion and contraction of austenite, respectively, upon continuous heating and subsequent cooling. The plateau of EF corresponds to $\gamma \rightarrow \alpha$ reaction. After completion of $\gamma \rightarrow \alpha$ transformation, the length of the sample decreases continuously down to room temperature due to normal thermal shrinkage (indicated by FG). Two plateaus of BC and EF revealing the $\alpha \leftrightarrow \gamma$ transformation account for the presence of the corresponding peaks A and B in Figure 4.15b. It can also be noted that the coefficient of thermal expansion of austenite is greater than that of ferrite/martensite.

Very similar features are observed in K8 (Figure 4.16). However, the segments BC and EF, which are indicative of $\alpha \leftrightarrow \gamma$ transformation, are characterized by a much steeper slope. This becomes clearly apparent by examining two sharp and narrow peaks marked by A and B in Figure 4.16b.

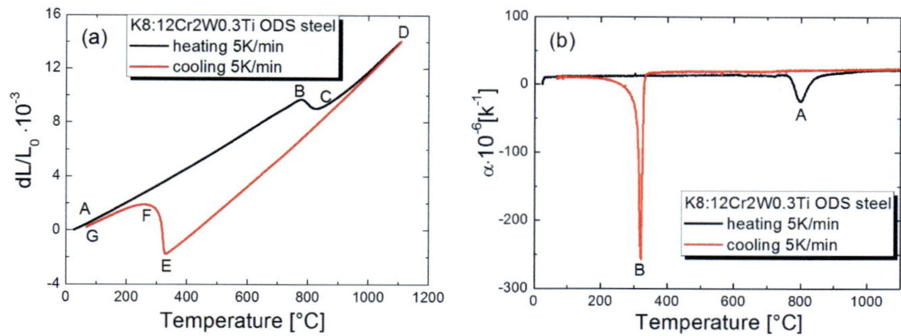

Figure 4.16: (a) Measured length change of 12Cr2W0.3Ti ODS steel during continuous heating and continuous cooling (5 K/min) and (b) Linear coefficient of thermal expansion α as a function of temperature. Initial length of the specimen is 25 mm.

The critical transformation temperatures are determined approximately by the deviation from tangential line in dilatometer curve and summarized in Table 4.3. It should be emphasized that the turning point is obtained from the two maxima in Figure 4.15b and Figure 4.16b. K15 exhibits the highest transformation temperatures, 458 to 414.9 °C for $\gamma \rightarrow \alpha$ reaction and 794.8 to 876 °C for $\alpha \rightarrow \gamma$ reaction, respectively. For 12Cr ODS steels, a broad transformation temperature range is observed in K7 which can be deduced from the smooth plateau in Figure 4.15a. In contrast, $\alpha \leftrightarrow \gamma$ transformation occurs in a narrow temperature range for K8.

Table 4.3: The critical transformation temperatures for 12Cr2W steels.

	M_s [°C]	M_t [°C]	M_f [°C]	A_s [°C]	A_t [°C]	A_f [°C]
K15	458.0	444.9	414.9	794.8	825.7	876.0
K7	396.8	356.1	320.3	757.5	780.4	814.6
K8	326.1	318.9	305.4	780.8	800.3	829.9

M_s: martensite start temperature; M_t: turning point; M_f: martensite finish temperature

A_s: austenite start temperature; A_t: turning point; A_f: austenite finish temperature

According to the critical temperatures listed above, the temperature of tempering treatment was chosen as 750 °C for 12Cr2W steels as shown in Table 3.2 to avoid phase transformation and obtain the good combination of strength and toughness. With respect to the microstructure, TEM results suggest that no evident difference is introduced by heat treatment for 12Cr ODS steels. The complex microstructure and high dislocation densities make it challenging for the quantitative microstructure analysis.

4.1.2.4 The influence of thermomechanical processing

In order to gain a full view of the grain morphology after TMP, 3 different cross-sectional planes (which are defined in Figure 4.17a) have been examined for the ODS steel K9 by both OM and SEM. As shown in Figure 4.17, the three passes cross rolling results in pronounced microstructure anisotropy. A fine elongated bamboo-like grain structure is observed in RD and TD directions as expected since cross rolling involves two perpendicular rolling axes. Although more deformation is introduced into RD direction than TD direction, both RD-ND and TD-ND planes have similar fine elongated grains. In contrast, the microstructure of RD-TD section consists of pancake shaped grains similar to the alloy with the same composition after HIP+HT. It seems that it is impossible to eliminate the second-phase precipitates by TMP. TMP does not lead to remarkable change in the shape and size of spherical precipitates. However, these precipitates align preferably along the deformation directions.

Figure 4.18 presents a close look of the microstructure after TMP. The grains vary from 0.4 to 0.6 µm in width and they are elongated with a factor of 4:1 along RD direction. The pancake shaped grains range from several hundred nanometers to about 2 µm in diameter. The RD-ND section exhibits a darker contrast suggesting a higher dislocation density.

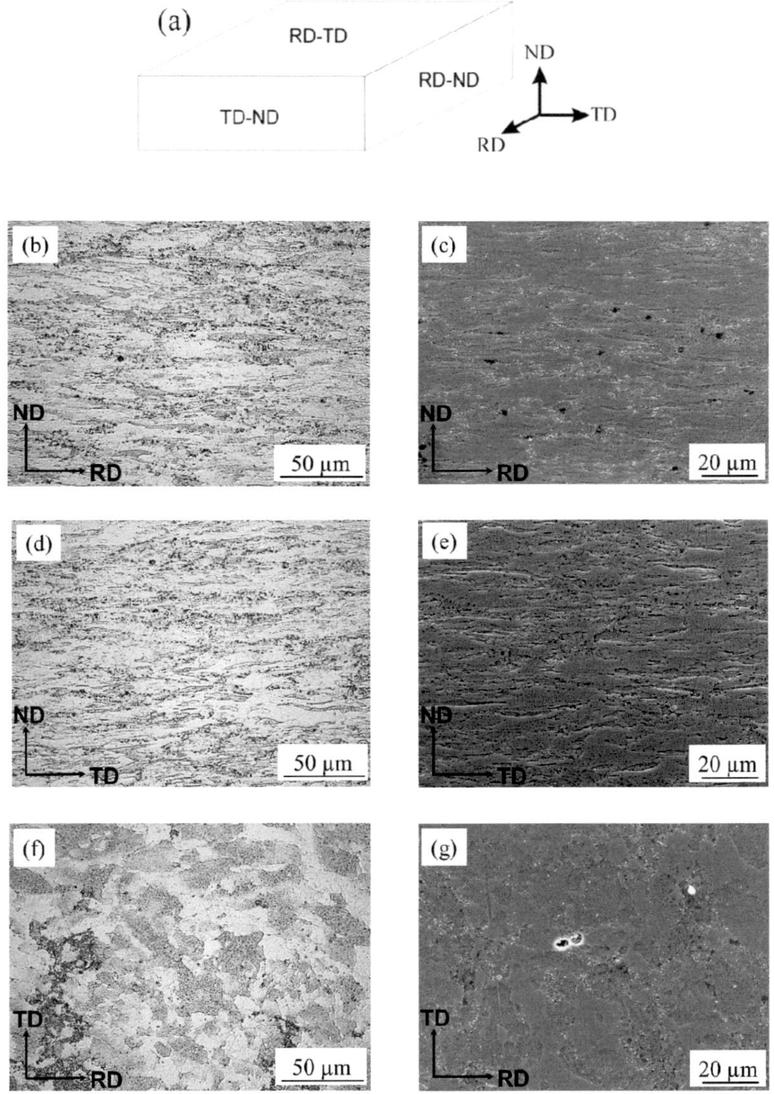

Figure 4.17: (a) Schematic drawing of sample orientations and microstructure evolution for 13.5Cr1.1W0.3Ti ODS steel K9 during TMP: optical micrographs (b) (d) (f) and SEM micrographs (c) (e) (g), respectively.

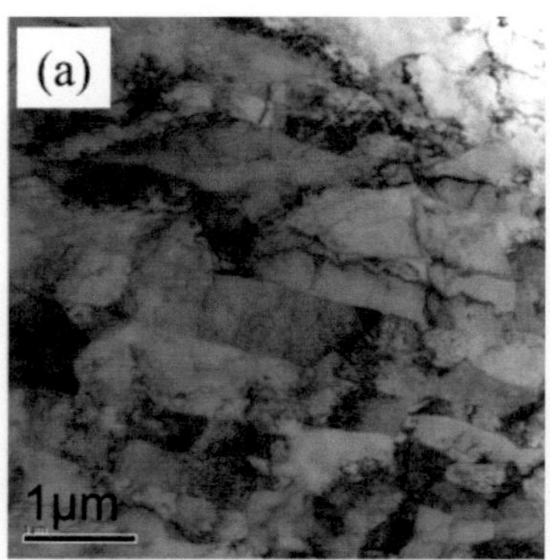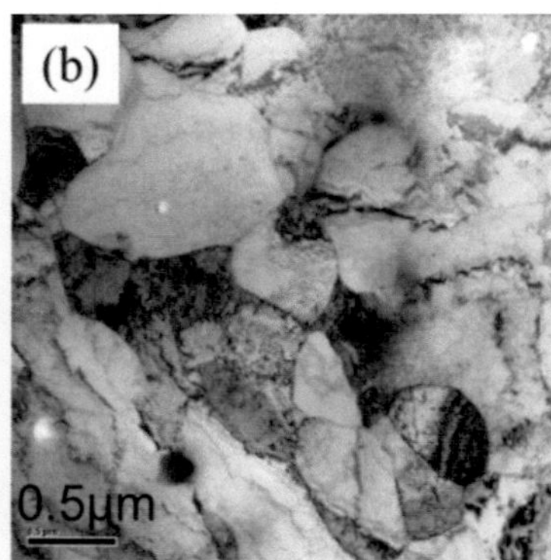

Figure 4.18: TEM micrographs for 13.5Cr1.1W0.3Ti ODS steel K9: (a) RD-ND section; (b) RD-TD section

EBSD scan was performed on a large area of 300×440 μm^2 from the RD-ND section after cross rolling and subsequent annealing with a step size of 200 nm. High resolution EBSD reveals important details of the microstructure. Figure 4.19a displays pronounced elongated grains along RD which is in consistent with previous results and blurred fine equiaxed grains where the subgrain boundaries cannot be resolved. Furthermore, Figure 4.19b shows the details of the grain structure in an enlarged view of the area marked by the rectangular in Figure 4.19a. High angle boundaries (above 15°) are marked with black lines while low angle boundaries (2–15°) are marked with white lines. The misorientation of some subgrain boundaries cannot be resolved by EBSD, especially those below 2°. Low angle grain boundaries are commonly found within the coarse elongated grains while massive nanoscale equiaxed grains are clearly visible, mainly boarded with high angle grains boundaries (HAGB).

The grain size distribution and the point-to-point misorientation profile are shown in Figure 4.20. The average grain size is 1.42 µm and the average misorientation angle is 27.08°. The grain size varies in a quite broad range from 0.5 to 10 µm and a bimodal grain size distribution is found with two maxima at about 2 μm and 8 μm. The misorientation profile reveals a large fraction (36%) of low angle grain boundaries (2–15°) resulting from the recovery process.

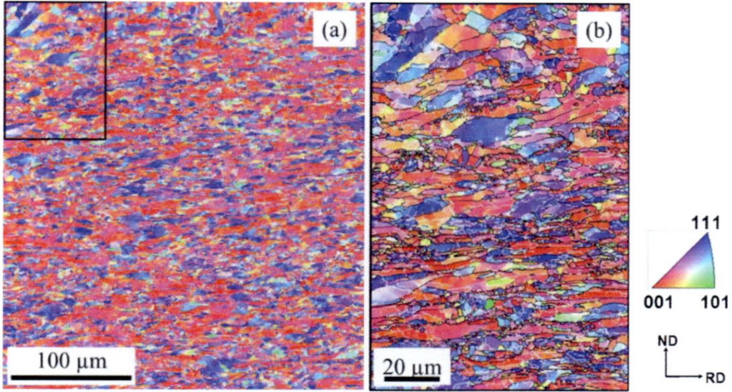

Figure 4.19: High resolution orientation map (a) and a magnified orientation map (b) from RD-ND section of K9. High angle boundaries (above 15°) are marked with black lines while low angle boundaries (2–15°) are marked with white lines.

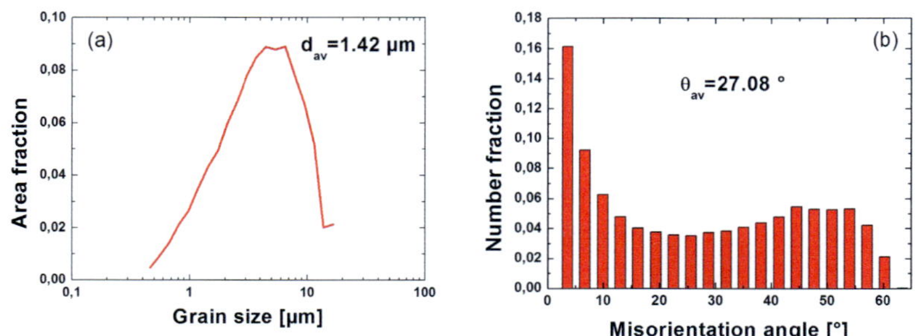

Figure 4.20: Grain size distribution (a) and misorientation angle distribution (b) of RD-ND section for 13.5Cr1.1W ODS steel K9 after TMP.

The residual strain distribution associated with the inhomogeneous microstructure was estimated by Kernel Average Misorientation (KAM). Figure 4.21 displays an orientation map and corresponding color-code KAM map of K9. High KAM values (3–5) are found to be associated with fine equiaxed grains, indicating a very high dislocation density. The lowest KAM value (0–1) corresponds to 10–20 µm coarse equiaxed grains softened by recrystallization. The recovered region is characterized by intermediate KAM values.

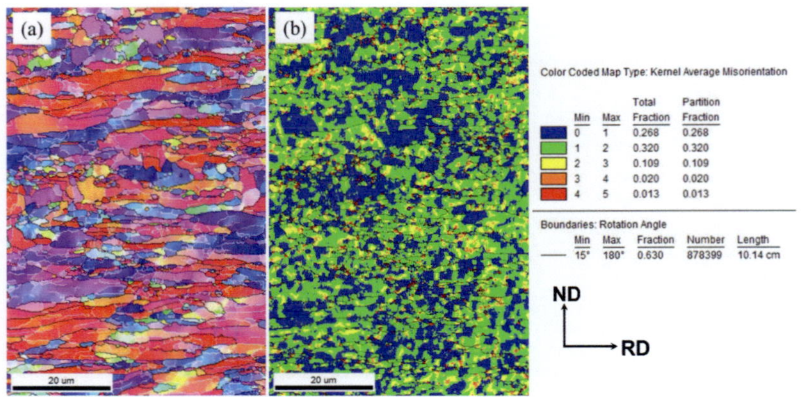

Figure 4.21: Orientation map (a) and corresponding Kernel average misorientation (KAM) map (b) of RD-ND section for 13.5Cr1.1W0.3Ti ODS steel K9 after TMP.

Low-carbon ferritic steels tend to develop strong fiber textures during cold rolling and further recrystallization annealing. The most relevant fibers are the so-called α-fiber (referred to all orientations with a common <110> axis along the rolling direction RD) and the γ-fiber (orientations with a common <111> axis parallel to the normal direction ND) [83]. The orientation of the grains are indicated in the respective {001}, {011}, and {111} discrete pole figures as shown in Figure 4.22a. They can be distinguished more accurately in the corresponding orientation distribution function (ODF) in sections with $\phi2 = 0°$ and $\phi2 = 45°$ (Bunge notation), as shown in Figure 4.22b. The developed texture has a sharp rotated-cube {001} <110> component. The γ-fiber, usually helpful for the improvement of ductility, is not present.

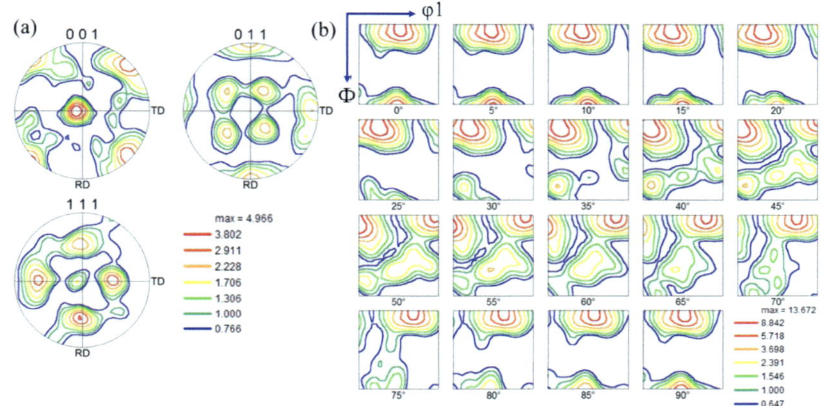

Figure 4.22: pole figures (PF) (a) and Orientation distribution function (ODF) plots (b) of 13.5Cr ODS steel K9 corresponding to orientation map in Figure 4.19a.

4.1.3 Composition evolution during fabrication

Table 4.4 shows the chemical composition of three base alloys. The concentration of main alloying elements, such as Cr and W, is very close to the nominal composition. Although appropriate clean steel making technologies have been applied, the achieved compositions deviate in some cases. The most notable feature is the unexpected high concentration of the radiologically undesired element Ni. In all these base alloys, K15 has the highest level of 1.69%. The undesired impurities can be explained by pick-up from small residues left in the equipment used for other steel compositions. Other elements, for instance, C, N, P and S are found to be within the acceptable range. The base powders especially in K15 represent higher values of O concentration compared to that in EUROFER 97.

Table 4.5 compares the evolution of the chemical composition of 13.5Cr2W ODS alloy during the fabrication process. The non-ODS alloy K12 was consolidated using the base powder by HIP. The composition of the as-hipped material remains approximately constant in comparison to the base powder indicating that HIP will not introduce an evident contamination. ODS steels were produced by mechanical alloying of the base powder with the addition of 0.3% yttria and different TiH_2 followed by HIP and subsequent heat treatment.

It is well known that the contamination is a serious problem for MA. Ti and Y concentration is found to be 0.286% and 0.186% in the MA powder, respectively. As expected, the concentration of C and N increase by an order of magnitude during MA which can be accounted for the wear of the milling medium and container. After MA, the total concentration of O is 0.09% and 0.06% O is introduced by the addition of 0.3% Y_2O_3. This suggests that the highly pure H_2 using as milling atmosphere is an effective measure to reduce O contamination. After HIP, O concentration increases further by 0.08% which is

Table 4.4: Chemical compositions of three base alloys.

	K12 (13.5Cr2W)	K15 (12Cr2W)	K13 (13.5Cr1.1W0.3Ti)
(A) Main alloying elements (wt. %)			
Cr	13.87	11.73	13.98
W	2.04	2.01	1.25
Ti	0.0125	0.0007	0.298
C	0.007	<0.005	0.011
N	0.005	<0.005	0.016
O	0.031	0.098	0.038
P	<0.01	<0.01	<0.01
S	<0.002	0.003	<0.002
(B) Undesired elements (wt. %)			
Al	0.0045	<0.0005	0.0168
Co	<0.005	<0.005	<0.005
Mn	0.003	0.002	0.006
Mo	<0.0005	<0.0005	<0.0005
Nb	0.00184	0.00511	0.00335
Ni	0.74	1.63	0.205
Si	0.041	0.026	0.072
V	0.053	0.115	0.015

the main reason for the occurrence of numerous coarse oxide precipitates. The impurities in the as-hipped ODS alloy increase slightly except Co and Al.

Table 4.5: The comparison of chemical composition for 13.5Cr2W ODS alloys during the fabrication process.

	K12 (Raw)	K12 (HIP)	K5 (MA)	K5 (HIP)
	(A) Main alloying elements (wt. %)			
Cr	13.87	13.57	13.3	13.68
W	2.04	1.94	1.92	1.835
Ti	0.0125	0.0042	0.286	0.270
Y	-	-	0.186	0.237
C	0.007	0.004	0.0466	0.065
N	0.005	<0.002	0.024	0.063
O	0.031	0.032	0.09	0.169
P	<0.01	<0.01	<0.01	<0.01
S	<0.002	<0.002	<0.002	0.0028
	(B) Undesired elements (wt. %)			
Al	0.0045	<0.005	0.0059	0.0108
Co	<0.005	<0.005	0.0848	0.401
Mn	0.003	0.0029	0.023	0.0208
Mo	<0.0005	<0.0005	0.0013	0.0061
Nb	0.00184	<0.001	0.002	<0.005
Ni	0.74	0.759	0.725	0.746
Si	0.041	0.035	0.05	0.067
V	0.053	0.052	0.0548	0.0519

The chemical composition of 13.5Cr1.1W ODS steels is listed in Table 4.6. TMP leads to a remarkable increase in C concentration. ODS steel after TMP exhibits lower levels of N, O and Co than the as-hipped material. This result suggests that TMP does not lead to additional contamination.

Table 4.6: The comparison of chemical composition for 13.5Cr1.1W0.3Ti ODS alloys during the fabrication process.

	K13 (Raw)	K13 (HIP)	K1 (MA+HIP)	K9 (TMP)
(A) Main alloying elements (wt. %)				
Cr	13.98 [13.5]	14.38	14.1	13.5
W	1.25 [1.1]	1.186	1.151	1.20
Ti	0.298 [0.3]	0.295	0.288	0.284
Y	-	-	0.202	0.17
C	0.0168	0.015	0.012	0.0358
N	0.011	0.018	0.024	0.019
O	0.038	0.045	0.176	0.131
P	<0.01	<0.01	<0.01	<0.01
S	<0.002	0.0015	0.0031	0.0011
(B) Undesired elements (wt. %)				
Al	0.015	0.0175	0.0185	0.0108
Co	<0.005	<0.005	0.294	0.0965
Mn	<0.0005	0.0054	0.0229	0.0019
Mo	0.0168	<0.0005	0.0128	0.0032
Nb	0.205	<0.005	<0.005	0.0037
Ni	0.205	0.225	0.249	0.251
Si	0.072	0.070	0.091	0.084
V	0.00335	0.0165	0.0172	0.0205

Table 4.7 reveals the composition evolution for 12Cr2W ODS alloying during the fabrication process. High Ni concentration from the base powder remains unchanged during MA and HIP. Ni is known as austenite stabilizer and plays a significant role in the formation of F/M structure. The concentration of C, N and O lies in the similar level compared to 13.5Cr ODS steels. Furthermore, the variations in C, N, O concentration for two 12Cr ODS steels result in different dilatometric curves and thus relative ferrite/martensite ratio.

Table 4.7: The chemical composition for 12Cr2W ODS alloys during the fabrication process.

	K15 (Raw)	K15 (HIP)	K7 (HIP)	K8 (HIP)
(A) Main alloying elements (wt. %)				
Cr	11.73 [12]	11.86	11.75	11.95
W	2.01 [2]	2.03	2.00	1.99
Ti	0.0007	0.0033	0.0176	0.276
Y	<0.0005	<0.0005	0.209	0.210
C	<0.005	<0.005	0.0287	0.009
N	<0.005	0.0061	0.0077	0.0055
O	0.098	0.0956	0.189	0.158
P	<0.01	0.006	0.009	0.007
S	0.003	0.003	0.003	0.004
(B) Undesired elements (wt. %)				
Al	<0.0005	<0.0005	<0.0005	0.0007
Co	<0.005	0.005	0.0709	0.0915
Mn	0.002	0.0013	0.0069	0.0075
Mo	<0.0005	<0.0005	0.0013	0.0023
Nb	0.00511	0.0056	0.006	0.0056
Ni	1.63	1.68	1.66	1.60
Si	0.026	<0.01	0.036	0.038
V	0.115	0.107	0.106	0.117

4.2 Oxide particles

Although it has been known for a decade that extraordinary mechanical properties of ODS steels originate from highly stabilized oxide dispersoids with a size smaller than 10 nm, the characteristics of these dispersoids (the size distribution, the chemical composition and the structure) remain as one of the most important scientific issues in nuclear materials research. The coarse precipitates in the range of 0.1–1 μm have a minor influence on the tensile properties. Under high strain rate deformation, however, they can considerably degrade toughness properties or reduce the lifetime during thermal or isothermal low cycle fatigue testing. An in depth characterization of these nanoscale dispersoids and coarse precipitates by TEM is presented in this section.

4.2.1 Nano dispersoids

The ODS dispersoids are visible as round spots on the bright field TEM micrograph with a contrast similar to that of the matrix, as shown in Figure 4.23. The observed low contrast of ODS dispersoids requires an underfocused regime to make them distinguishable from the matrix background. This suggests that the orientation of ODS dispersoids should be similar to that of the matrix. Evaluation of numerous TEM micrographs shows that ODS dispersoids are homogeneously distributed within the grains. The dispersoids with relatively large size are often found along the grain boundaries.

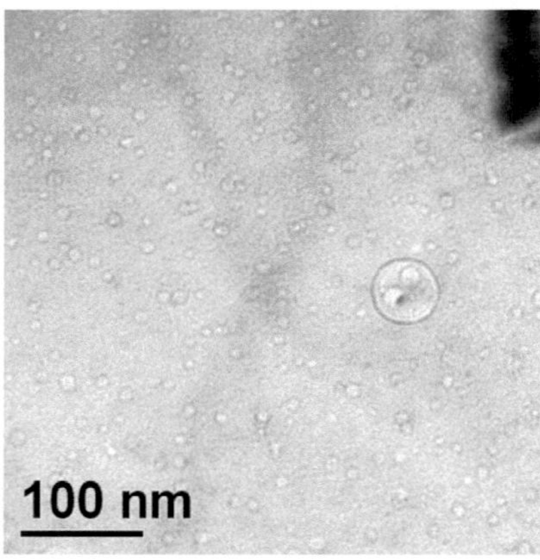

Figure 4.23: TEM bright field image of ODS particles for 13.5Cr2W ODS steel K2.

The corresponding size distribution histograms and mean particle size of ODS dispersoids obtained from numerous micrographs for samples with different Ti contents are shown in Figure 4.24 and Figure 4.25. The evaluation was on the basis of 166, 158, 235, 168 dispersoids, respectively. For the Ti-free sample, the ODS dispersoids distribute over a wide size range of 2 to 20 nm. The Ti addition shows striking influence on the refinement of the dispersoids. The mean particle size for all Ti-containing drops to less than 6 nm in diameter. The best refinement result is achieved by shifting the average diameter from 11.48 nm for the non-Ti ODS alloy to 3.73 nm with the addition of 0.3% Ti. Further increase in Ti content does not seem very effective in particle refinement. However, two issues must be taken into account concerning the particle size distribution. Firstly, constrained by a detection limit, it is impossible to obtain reliable overall particle size statistics especially for these ultrafine nanoclusters smaller than 2 nm in diameter. Secondly, nano grains were clearly visible in all Ti-containing ODS alloys with a higher dislocation density. The contrast produced by closely located dislocation lines overlaps with that from ODS dispersoids. Therefore ODS particles are often invisible within nano grains. Only relative coarse particles, 4–10 nm in diameter, are clearly visible along grain boundary. In this work, the micrographs for particle size evaluation are mainly taken from the dislocation-free coarse grain region.

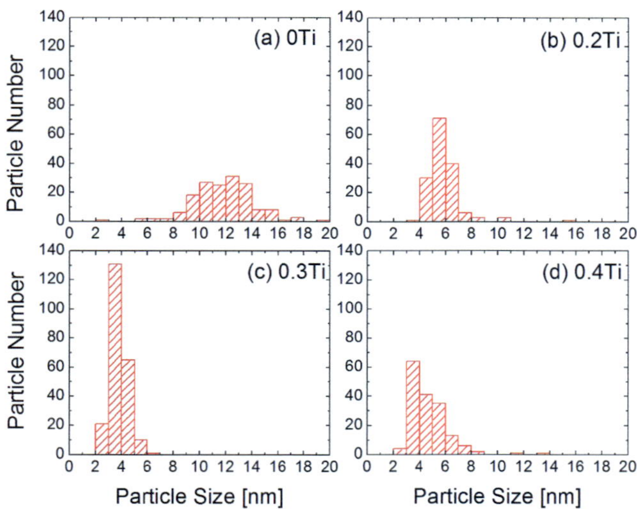

Figure 4.24: ODS dispersoid size distribution histograms obtained from as-hipped 13.5Cr2W ODS alloys (a) 0Ti (K2) (b) 0.2Ti (K4) (c) 0.3Ti (K5) (4) 0.4Ti (K6).

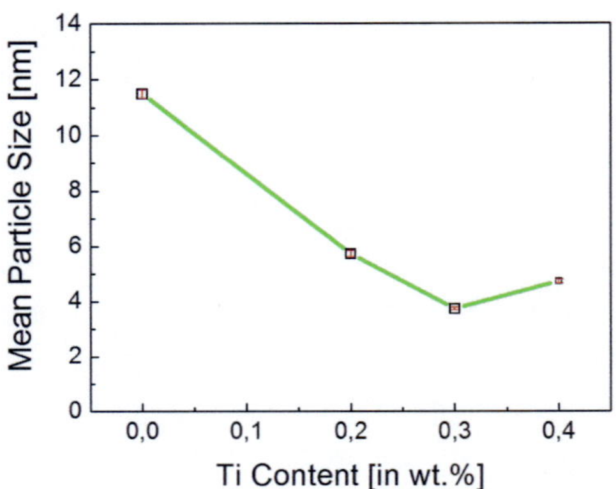

Figure 4.25: Ti dependence of mean particle size for different hipped ODS alloys.

Figure 4.26 shows the chemical composition of a particle with a size of 20 nm in 13.5Cr2W ODS steel without Ti (K2). The presence of Y is well confirmed by the EDX profile and spectrum. EELS profile and spectrum clearly reveals the existence of O. It is remarkable that the intensity of O increases dramatically within the particle by EELS line scan. An evident O-K edge is observed in EELS spectrum after background subtraction. The line scan allows one to distinguish the influence from the matrix (Fe) and main alloying element (Cr). Within the particle, the intensities of Fe and Cr decrease by a factor of 1.75 and 1.3 respectively. All these results may be considered as an indication of the formation of Y oxide. However, it is not sufficient to prove whether high Cr intensity attributes to the matrix effect or to the formation of Y-Cr-O oxides.

Along with HAADF micrograph, EDX elemental maps provide chemical composition of ODS dispersoids over large areas. The HAADF micrograph (Figure 4.27a) only shows ODS dispersoids that are relatively large and preferably orientated to the electron beam axis with an adequate contrast. The darker round dots in Fe map correlate well with the bright contrast in Y map. Only one brighter dot with a size of 100 nm is clearly visible in Cr map due to the poor signal-to-noise ratio and/or smaller decrease in Cr intensity within the ODS dispersoids. At upper left, a Y rich area exists in the Cr-enriched precipitate of 100 nm in size. This indicates the presence of Y-O and Cr-O complex oxide.

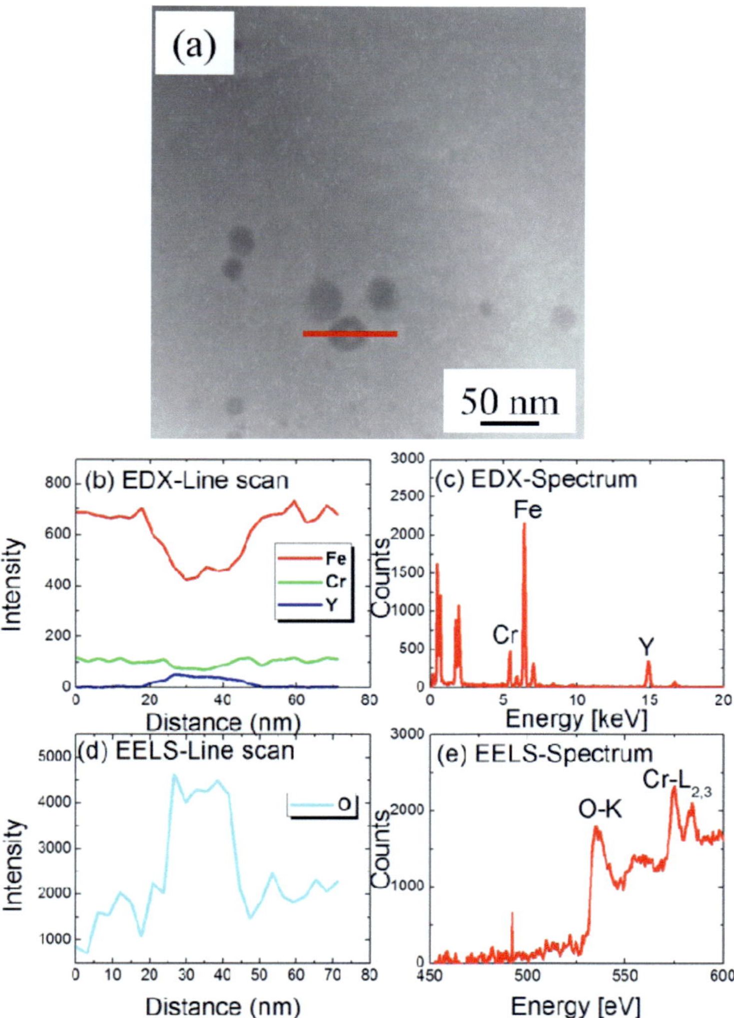

Figure 4.26: (a) HAADF image of Y oxide with marked line for scanning experiments (b) EDX and (d) EELS relative intensities along the scanning line; (c) EDX and (e) EELS spectra acquired from middle of the same particle in 13.5Cr2W ODS steel K2.

The addition of Ti is not only beneficial to refine the particle size and increase the particle number density, it can also lead to significant changes in the chemical composition of ODS dispersoids. HAADF micrograph and EDX elemental distribution maps in 13.5Cr2W0.4Ti ODS steel are presented in Figure 4.28. The particles (indicated by green arrows) located at grain boundaries (indicated by red dashed lines) have a larger size, which is likely detectable. All these particles with Fe and Cr deficiency are imaged as dark spots on the Fe (Figure 4.28b) and Cr (Figure 4.28c) maps. The spatial distributions of Ti and Y on the elemental maps (Figure 4.28d and e) are in good agreement with Fe and Cr deficiency. The EDX elemental mapping clearly reveals the presence of Y-Ti-O oxides. However, it cannot exclude the existence of Y_2O_3 in Ti containing ODS alloys.

Y-Ti-O complex dispersoids play a dominate role in 0.4Ti ODS steel. Nevertheless, some of these dispersoids contained a non-negligible content of Al. Figure 4.29 reveals the formation of (Y, Ti, Al) oxide. There are two main sources for the detected Al. It can be attributed to the steel powder manufacturing process where Al is commonly used as deoxidizers to reduce the dissolved oxygen in the molten steel to a required level. In addition, MA is considered as an important process for powder contamination. Although only 0.01% Al is contained in 0.4Ti ODS alloy after HIP and HT, it can strongly affect the formation of Y-Ti-O dispersoids.

Figure 4.30 presents the spatial distribution and the chemical composition of the ODS dispersoids in 13.5Cr1.1W0.3Ti steel (K9) after TMP. These dispersoids are characterized as Y-Ti-O by EDX elemental maps which are similar to that in 13.5Cr2W0.4Ti ODS steel (Figure 4.28). After TMP, elongated grains along the RD and the obvious alignment of ODS dispersoids are observed. Some (Y-Ti-V) oxides are detected in K9, as shown in Figure 4.31. These dispersoids may result from a cross contamination during MA process.

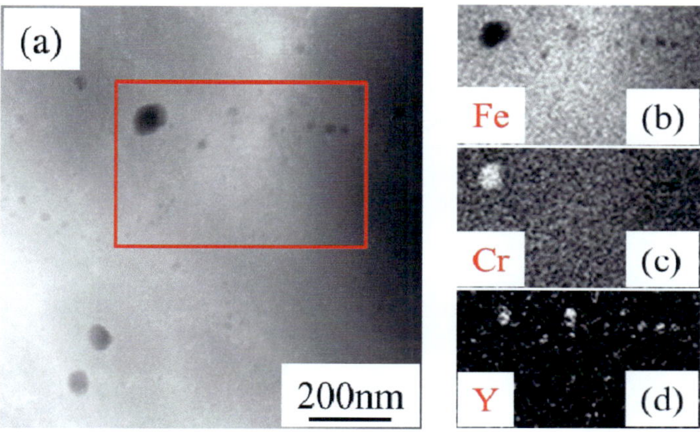

Figure 4.27: (a) A HAADF micrograph of area with ODS particles; (b) EDX elemental maps of Fe, Cr and Y in 13.5Cr2W0.4Ti ODS alloy K6.

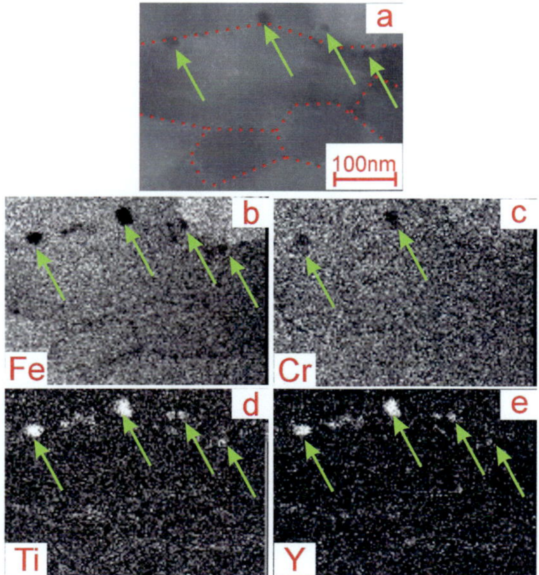

Figure 4.28: (a) HAADF micrograph and EDX elemental distribution maps (b) Fe, (c) Cr, (d) Ti and (e) Y of Y-Ti-O particles in 13.5Cr2W0.4Ti ODS steel.

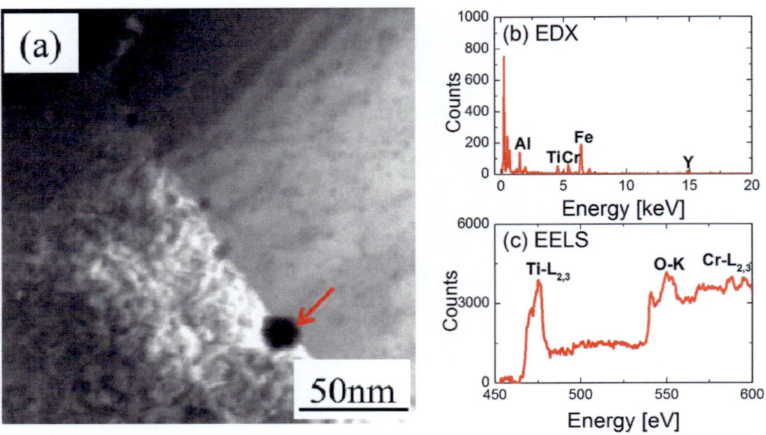

Figure 4.29: (a) A HAADF micrograph of ODS dispersoids in 13.5Cr2W0.4Ti ODS steel; (b) EDX and (c) EELS spectra acquired from the dispersoid indicated by the arrow.

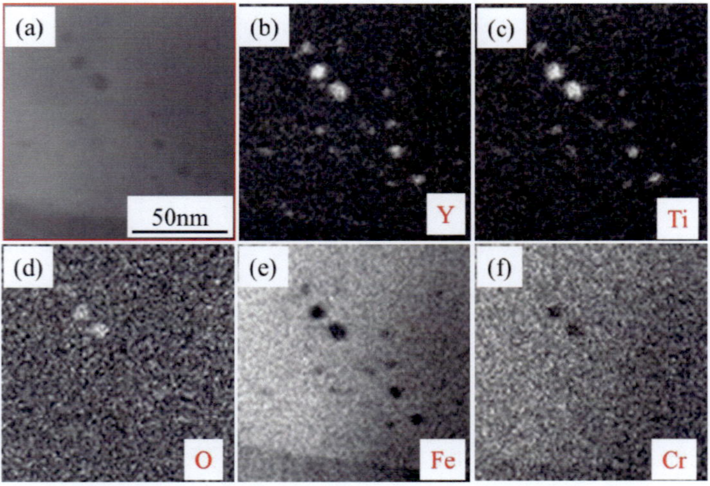

Figure 4.30: (a) HAADF micrograph and EDX elemental distribution maps (b) Y, (c) Ti, (d) O (e) Fe and (f) Cr of Y-Ti-O particles in 13.5Cr1.1W0.3Ti ODS steel after TMP.

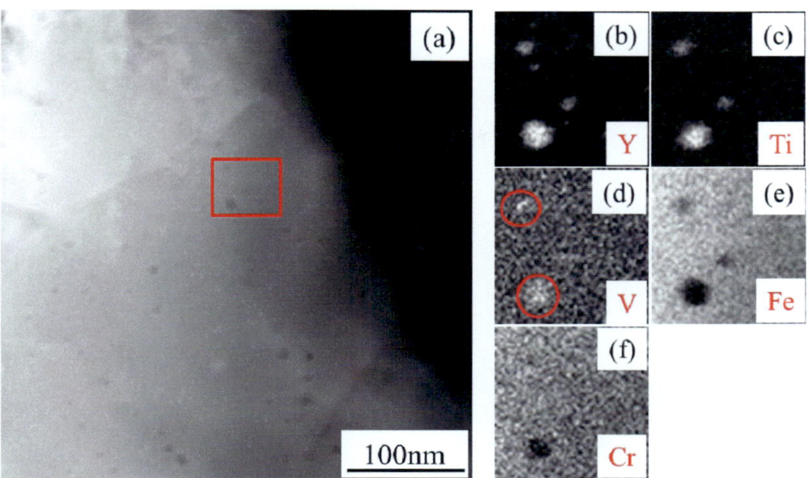

Figure 4.31: (a) HAADF micrograph and EDX elemental distribution maps (b) Y, (c) Ti, (d) O (e) Fe and (f) Cr of Y-Ti-O particles in 13.5Cr1.1W0.3Ti ODS steel K9 after TMP revealing the presence of (Y, Ti, V) oxides.

4.2.2 Submicro precipitates

As already shown in Figure 4.7, approximately 30–40% of the grain boundaries are covered by elongated coarse precipitates in all the as-hipped 13.5Cr2W ODS steels. The chemical nature of these precipitates was investigated by means of EDX and EELS analysis as shown in Figure 4.32. To exclude the influence of the surrounding matrix and the carbon from contamination, the measurements were performed mainly at precipitates located very close to the edge of the hole in TEM foil samples. EDX and EELS results led to the conclusion that these elongated precipitates were Cr carbides. Cr carbide exhibits brighter contrast compared to the surrounding matrix in Figure 4.32b. At first glance this result is contradictory to the basic principle of a HAADF image: higher Z elements with brighter contrast while lower Z elements with darker contrast. However, Cr carbides remain thicker than the matrix because Fe-matrix is prone to dissolve during electrochemical polishing. Since the base alloy has a very low carbon (C) content of 0.005%, the C necessary to form these massive precipitates can be mainly attributed to a pick-up during the mechanical alloying due to the wear of the high C containing milling balls. Similar precipitates were observed in as-hipped EUROFER ODS, which had a C content of about 0.08%. They have been considered as the main reason for the degradation in tensile ductility, impact properties as well as corrosion resistance [68].

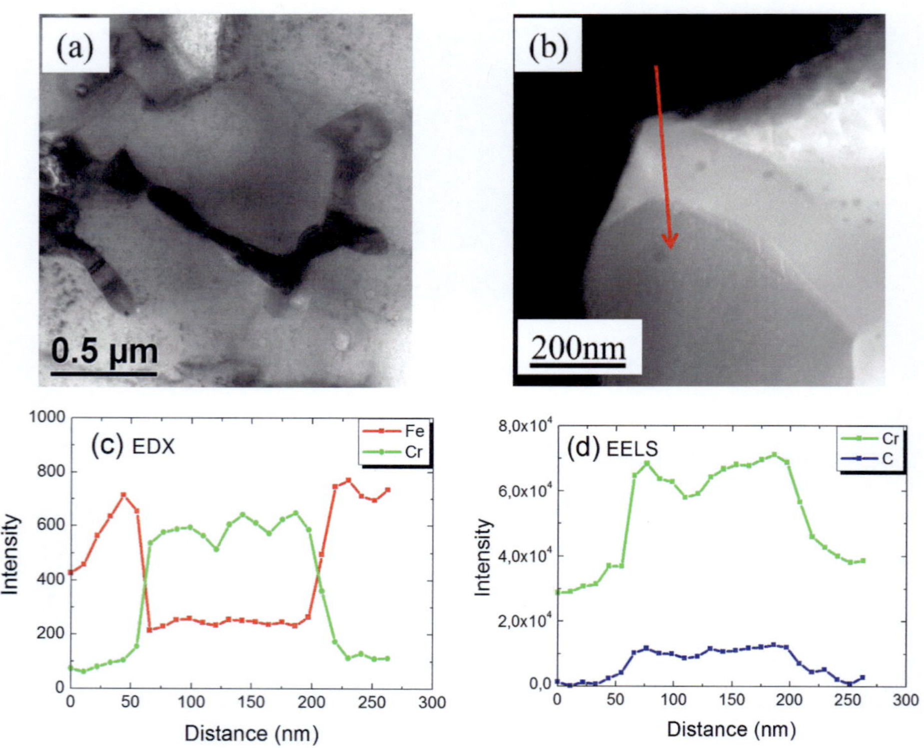

Figure 4.32: (a) BFTEM micrograph of Cr carbides (b) HAADF image of Cr carbide with marked line for scanning experiments (c) and (d) show the intensity of EDX and EELS profiles along marked scanning line in as-hipped 13.5Cr2W ODS steel K2.

Besides the elongated Cr carbides, the chemical composition of spherical precipitates with a size from 100 to 300 nm in diameter was revealed by EDX elemental mapping over large area. The HAADF micrograph as well as elemental mapping images acquired using Fe, Cr, O and Y is presented in Figure 4.33. For the 0% Ti alloy, the precipitates are visible as round spots on Figure 4.33a, with a slight darker contrast. Elemental mapping of Fe (Figure 4.33b) clearly reveals regions of local Fe deficiency owing to the superior signal/noise ratio. The Cr image (Fig. 4.33c) and O image exhibit patterns of local Cr and O enrichment which perfectly correlate with the Fe deficiency map. On the Y element map, only several ODS particles can be detected, which are marked with red circles. These findings suggest the coexistence of Cr oxides and Y-Cr-O particles. However, Cr-O/Y-Cr-O complex oxides are occasionally observed in the 0% Ti alloy and coarse Cr oxides exhibit a much smaller number density and volume fraction compared to Y oxides.

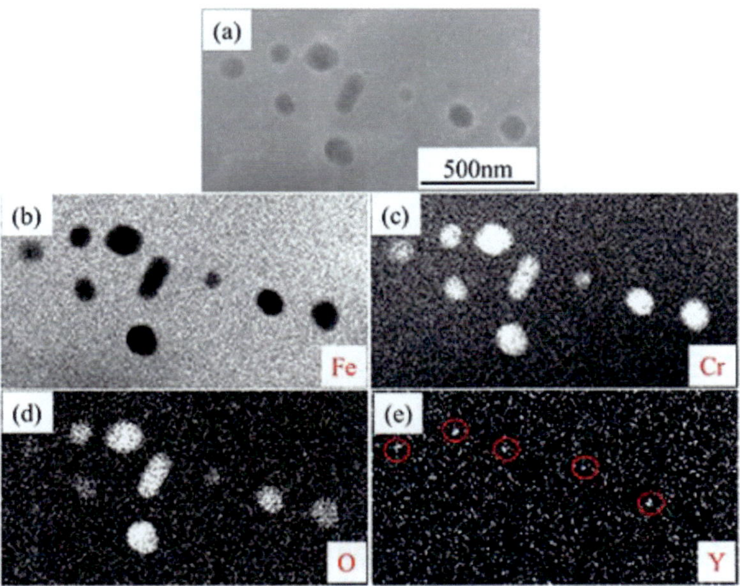

Figure 4.33: (a) HAADF image and EDX elemental distribution maps (b) Fe, (c) Cr, (d) O and (e) Y of Cr-O/Y-Cr-O particles in 13.5Cr2W ODS alloy.

Furthermore, numerous EDX and EELS investigations were performed on the similar precipitates for Ti-containing ODS steels, as shown in Figure 4.34. The precipitates in 0.4Ti ODS steels are similar to that in Figure 4.33 regarding the size, shape and contrast. The Cr and Ti maps, however, reveal a different chemical nature. These precipitates in 0.4Ti ODS steel were proven to Ti oxides rather than Cr oxides despite much lower content of Ti in comparison to Cr. The presence of O K-edge was confirmed by EELS measurement, which is not shown here. Obviously the addition of Ti changes the chemical equilibrium from the formation of Cr oxides towards Ti oxides because of a greater affinity of Ti to O.

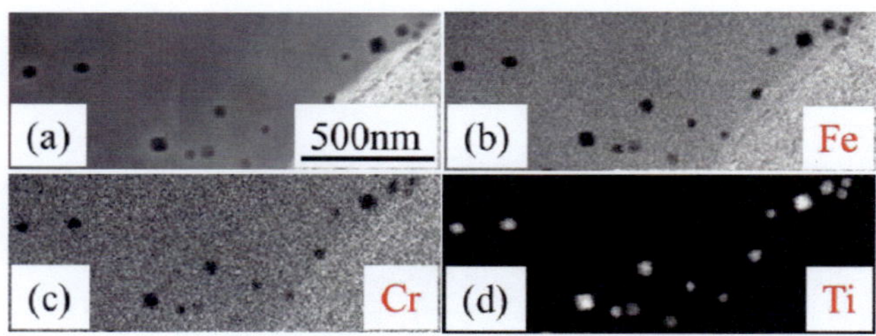

Figure 4.34: (a) HAADF image and EDX elemental distribution maps (b) Fe, (c) Cr, (d) Ti of Ti oxides in 13.5Cr2W0.4Ti ODS alloy K6.

Non-negligible Al was found within the ODS dispersoids (Figure 4.29). Al was detected within Cr oxide in Ti-free ODS steel (Figure 4.35) and Ti oxide in 0.3Ti ODS steel (Figure 4.36). Figure 4.35 presents the line scanning result of a precipitate with a size of 120 nm in diameter. In some cases, remarkable Al signal was detected within the Cr oxide precipitate. The contrast of the precipitate indicates that the precipitate consists of only one (Cr, Al) -O phase.

Complex precipitate composed of two or even more parts with different contrasts are found in 13.5Cr1.1W0.3Ti ODS steel (Figure 4.36). Different parts of the precipitate, indicated by the red dashed curve, may exist side by side. Simultaneous enrichment of Al and O was found within the part with relatively brighter contrast on the top right corner. When the scanning continues, the intensity of Al almost decreases to zero whereas a dramatic increase in Ti intensity is clearly visible. The highest O concentration is found at the interface of Al oxide and Ti oxide.

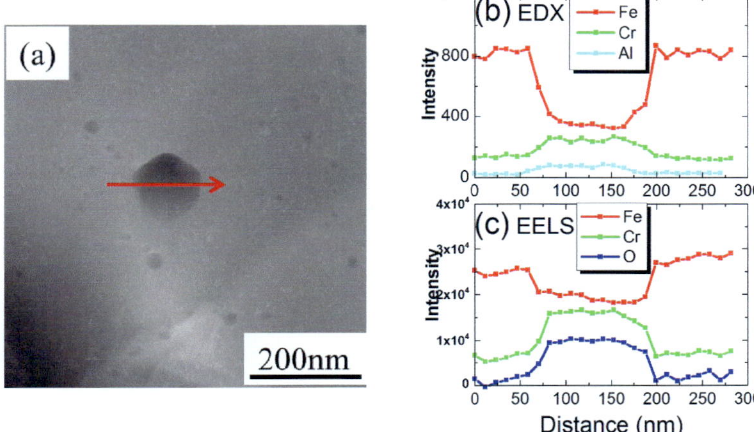

Figure 4.35: (a) HAADF image of (Cr, Al) oxide with marked line for scanning experiments (b) and (c) show the intensity of EDX and EELS profiles along marked scanning line in as-hipped 13.5Cr2W ODS steel K2.

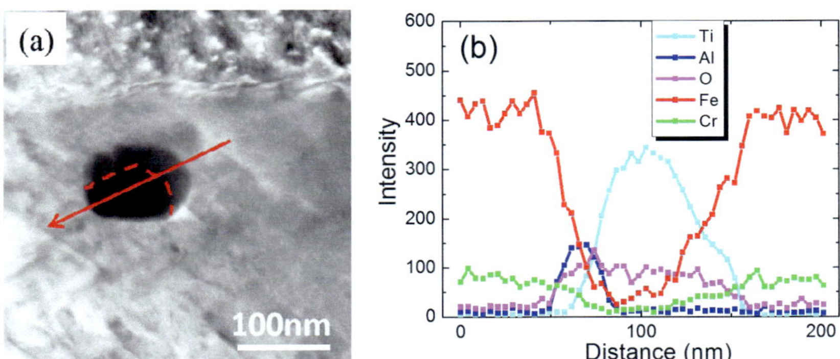

Figure 4.36: (a) HAADF image of complex precipitate Al oxide/ Ti oxide with marked line for scanning experiments (b) show the intensity of EDX profiles along marked scanning line in as-hipped 13.5Cr1.1W0.4Ti ODS steel K6.

In Figure 4.37, EDX maps of an area with a precipitate, 150 nm in diameter, are presented. The precipitate shows a darker contrast in the HAADF image as well as in Y, Fe and Cr maps and a brighter contrast in Ti and O maps. The interface of the precipitate and the matrix is clearly observed in the Fe and Cr maps, indicating a concentration gradient of Fe and Cr from the matrix to the precipitate.

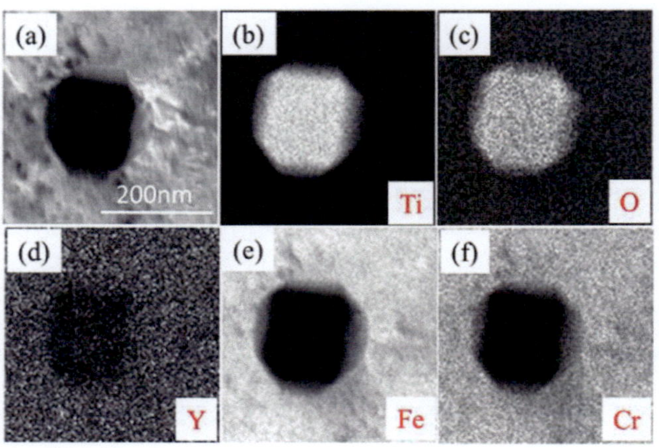

Figure 4.37: (a) HAADF image and EDX elemental distribution maps (b) Fe, (c) Cr, (d) O and (e) Y of Ti oxide precipitate in 13.5Cr1.1W0.3Ti ODS alloy K9.

4.3 XAFS

4.3.1 XANES

The XAFS investigations of ODS steels have been focused on the Y K-edge because Y is the major constituent of the nanoscale particles. The absorption edge, E_0 lying around 17.038 keV, does not interfere with the other elements in the steel (Fe: 7.112 keV; Cr: 5.989 keV; W: 69.525 keV; Ti: 4.966 keV) [84]. The K-edge in XANES, as indicated by arrow A in Figure 4.38, is sensitive to the oxidation state of Y. A higher valence of the absorber usually has an absorption edge shifting towards higher energy. The normalized XANES spectra for four MA powders with the two references, Y foil and Y_2O_3, are compared. Y_2O_3 features separated peaks (B and C) of higher intensities and a shift towards higher energy. For MA powders, the shoulder C above the main peak B is indistinct and all the spectra seem identical, which suggests that the entire Y in the various MA powders have similar chemical environments, and that the influence of Ti is trivial. The smeared-out peak C indicates a structure without long range ordering in nano Y_2O_3 particles [85]. The chemical shift of the MA powders lies between Y metal and Y_2O_3, implying that there is a portion of the Y in the metallic state.

In order to resolve the relative ratio of metallic Y over total Y, linear combination fits of the normalized XANES for MA powders were performed using Y metal and Y_2O_3. The fitting result is satisfactory, as shown in Figure 4.39a. An alternative model was tried by replacing Y_2O_3 with a compacted sample with 0.4% Ti. A good agreement is observed in Figure 4.39b and the relative ratio $Y_{dissolve}/Y_{total}$ in weight percentage can be obtained.

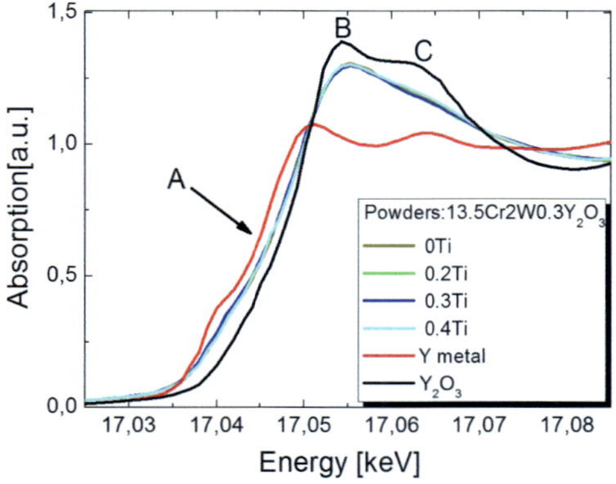

Figure 4.38: Y K-edge XANES spectra of the powder samples after MA together with the reference samples, metal Y and Y_2O_3, the absorption edge, the main peak and the shoulder at the post edge are marked by A, B and C, respectively.

The $Y_{dissolve}/Y_{total}$ ratio for all the MA powders is summarized in Table 4.8. The metallic Y component lies between 10.2% and 13.7% and varies slightly with the minor changes in the amount of Ti. The metallic Y component arises from either Fe-Y solid solution or pure Y. Because of the tiny difference of charge transfer for Y-Y and Y-Fe, their similar chemical shifts cannot be distinguished by XANES. Driven by high energy milling, part (not all) of the Y_2O_3 may be reduced in the reducing atmosphere (H_2). Another possibility is that atomic Y diffuses into the Fe structure and forms a substitutional alloy.

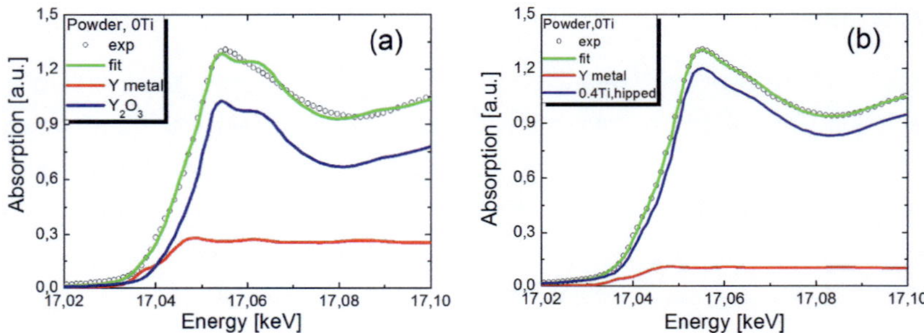

Figure 4.39: Y K-edge XANES of MA powders (0% Ti) and a linear fit by (a) Y metal and Y_2O_3, (b) by Y metal and compacted sample with 0.4% Ti.

Table 4.8: Relative fraction of the metallic Y component in the MA powders determined from the linear fit of XANES.

Composition	0Ti	0.2Ti	0.3Ti	0.4Ti
%	10.2	12.8	13.7	11.5

XANES spectra for the compacted samples are plotted in Figure 4.40. In contrast to the milled powders, the chemical shifts (marked by the arrow A) of the compacted samples are very close to Y_2O_3. A linear fit as employed for the steel powder samples does not give satisfying results. This implies that the dissolved Y and other alloying elements (Ti) reprecipitate from the steel during HIP. The compacted sample with 0.3% Ti exhibits the highest absorption peak B. And the 0.4% Ti sample shows an increased intensity in shoulder C, which suggests the presence of new Y oxide phases.

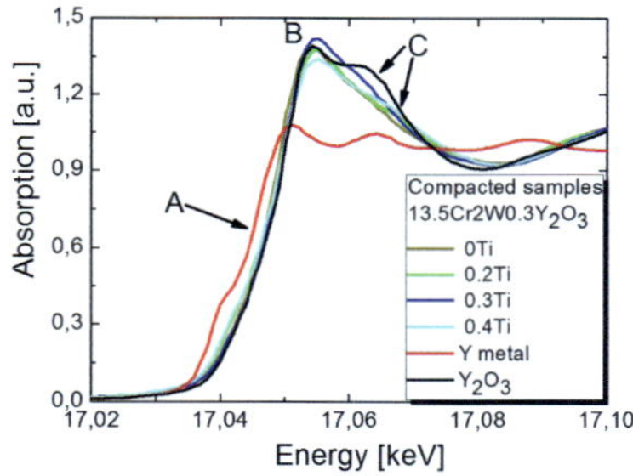

Figure 4.40: Y K-edge XANES of the compacted samples and the reference samples, Y and Y_2O_3, the absorption edge, the main peak and the shoulder at the post edge are marked by A, B and C, respectively.

4.3.2 EXAFS

The Fourier transform (FT) of the EXAFS functions of the powder samples and two references are shown in Figure 4.41. The primary feature of Figure 4.41 is two dominant peaks for pure Y_2O_3, which are assigned to the Y-O and Y-Y coordination shells. The first Y-Y shell in the Y metal has a larger R than that for the second Y-Y shell in the Y_2O_3. The MA powders are characterized by two similar peaks, close to Y_2O_3. The much smaller peak height of the Y-Y indicates that the Y_2O_3 in the MA powders is much more disordered because of severe plastic deformation. The small portion of metallic Y component is invisible because of their overlap with the second Y-Y shell of Y_2O_3. No evident difference is observed for the MA powders with differing amounts of Ti.

Figure 4.42 presents the theoretical fitting of the first two shells using the phase shift and backscattering amplitude extracted from the crystallographic structure of cubic Y_2O_3. The reasonable fitting results for all the powder samples are listed in Table 4.9. The local structure of cubic Y_2O_3 can be divided into three shells: Y-O with a bonding length R=2.26 Å, Y-Y1 with R=3.50–3.53 Å and Y-Y2 with R= 4.04–4.09 Å. The similarity of the FTs of the pure Y_2O_3 and the MA powders indicates that Y_2O_3 is the main constituent of Y even after 24 hours milling, which is consistent with the results of the XANES. However, the yttrium oxides after the MA are different from the initially added Y_2O_3 powders, with a decreased coordination number, especially for the second Y-Y shell. This can be explained by severe plastic deformation introduced by the high energy MA.

Large plastic deformation results in the generation of a stacking of various structural defects and refined crystallites, which destabilizes the ordered lattice and lead to the formation of more disordered (crystalline or amorphous) phases. It should be added that the fitted Y-Y1 shell may also contain a Y metallic component.

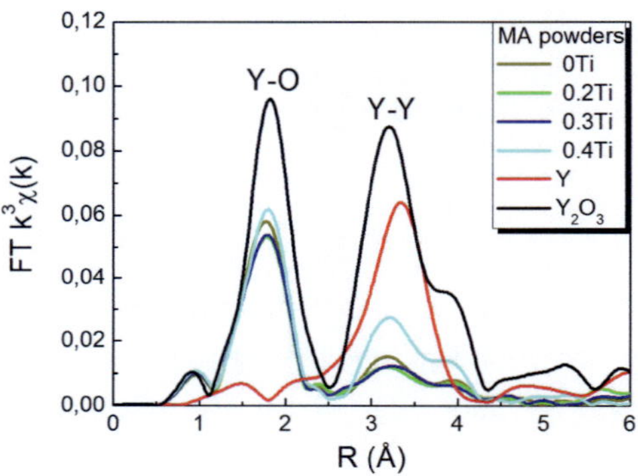

Figure 4.41: Fourier transform of EXAFS functions of MA steel powders and metal Y, Y_2O_3 references, phase shift was not corrected. The two peaks corresponding to Y-O and Y-Y are marked.

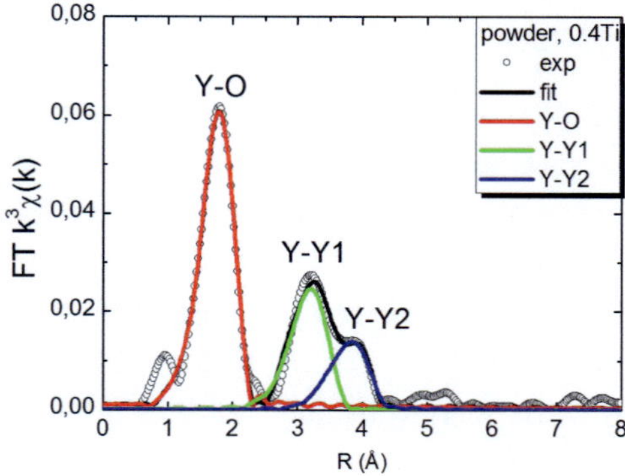

Figure 4.42: Fourier transform of EXAFS functions of MA powders with 0.4% Ti and fitting with three coordinations: Y-O, Y-Y1 and Y-Y2. Phase shift was not corrected.

Table 4.9: Fitted results of Y-O, Y-Y1, Y-Y2 shells for MA powders. The σ^2 for Y-Y1 and Y-Y2 were fixed. The data for c-Y_2O_3 were taken from the literature.

		c-Y_2O_3	0Ti	0.2Ti	0.3Ti	0.4Ti
	R(Å)	2.27	2.26	2.26	2.26	2.26
Y-O	CN	6	5.5	5.4	5.4	5.4
	σ^2	-	0.0089	0.0095	0.0094	0.0081
	R	3.52	3.52	3.5	3.52	3.53
Y-Y1	CN	6	1.9	1.5	1.5	3.4
	σ^2	-	0.01	0.01	0.01	0.01
	R	4.01	4.09	4.08	4.04	4.06
Y-Y2	CN	6	1.7	1.6	1.6	2.9
	σ^2	-	0.01	0.01	0.01	0.01

The fitting results for MA powder samples include the average coordination number (CN), bond length (R) and relative Debye-Waller factor (σ^2) at different coordination shells.

The FT of the EXAFS data for the compacted ODS samples is shown in Figure 4.43. After HIP, the local structure of Y shows a distinctive feature of Y oxides, but varying from one to another. It is noteworthy that different contents of Ti remarkably affect the local structure of Y. The sample with 0.3% Ti shows the largest Y-O peak. The Y-Y shell for the sample with 0% Ti splits into two peaks, whereas the samples containing Ti exhibit Y-Y shells similar to yttria.

In order to quantitatively interpret the trends observed in the FT, the first two peaks were fitted independently for the compacted samples. For the Ti-free compacted sample, the model Y_2O_3 did not have a good fit. Therefore, a new fitting model was selected with the complementary results obtained by TEM. An HRTEM study revealed the presence of cubic Y_2O_3 in the hipped EUROFER ODS alloy without Ti [38]. In spite of a relatively higher Gibbs energy of formation compared to Y_2O_3, a ternary oxide $YCrO_3$ coexists with Y_2O_3 in Fe-12Cr-0.4Y_2O_3 alloy after HIP and annealing [86]. Therefore, $YCrO_3$ was chosen as the model compound for EXAFS fitting for the Ti-free ODS sample. Y_2TiO_5 and $Y_2Ti_2O_7$ [42] were considered for the ODS samples containing Ti. The local structural parameters of some model compounds are listed in Table 4.10.

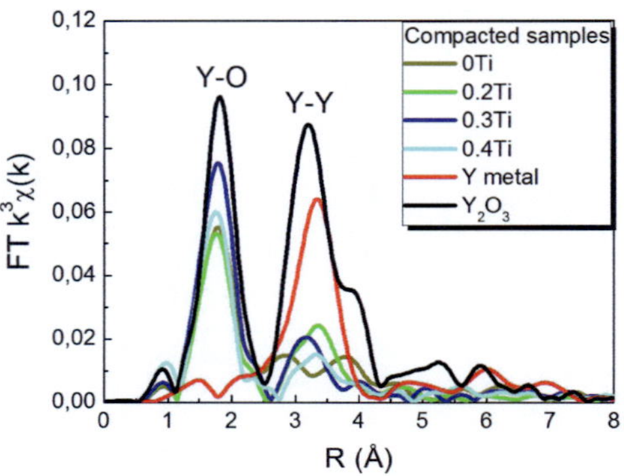

Figure 4.43: Fourier transform of EXAFS functions of compacted ODS samples with metal Y and Y_2O_3 references.

Table 4.10: Local structures around Y of some Y oxide compounds, including the coordination number, interatomic distance (Å) of each shell and space group.

Compound	First shell	Second shell	Third shell	Crystal structure	Space group	Reference
Y_2O_3	6O(2.28Å)	6Y(3.52Å)	6Y(4.00Å)	cubic	Ia-3	[87]
$YCrO_3$	8O(2.41Å)	6Cr(3.19Å)	6Y(3.83Å)	orthorhombic	Pbnm	[88]
$YFeO_3$	8O(2.43Å)	6Fe(3.21Å)	6Y(3.86Å)	orthorhombic	Pnma	[89]
Y_2TiO_5	7O(2.34Å)	4Ti(3.54Å) 3Ti(3.45Å)		orthorhombic	Pnma	[90]
$Y_2Ti_2O_7$	8O(2.43Å)	6Ti(3.57Å)	6Y(3.57Å)	cubic	Fd-3m	[91]

The interatomic distance R and the merit-of-fit with a model compound can be used for the identification of scattering atom species around a Y absorber. The fitting parameters for all compacted ODS samples are summarized in Table 4.11. Moreover, Fourier transform of EXAFS functions and data fitting of Ti free compacted ODS sample is shown in Figure 4.44. On the basis of the model $YCrO_3$, the first three coordination shells around the central Y absorber were fitted by Y-O, Y-Cr, and Y-Y coordinations, respectively. The first shell has an R of 2.28 Å, corresponding to a Y-O shell. The second shell, with an R of 3.22 Å, is likely to be either the Y-Fe shell with an R of 3.21 Å in $YFeO_3$ or the Y-Cr shell with an R of 3.19 Å in $YCrO_3$. The short distance excludes the possibility of Y-Y as in Y_2O_3. Based on the TEM observation which will be addressed later, the $YCrO_3$ model was used here. However, the R of the third shell, 4.01 Å, is larger than that in $YCrO_3$ (3.83 Å) but close to the R of 4.00 Å in cubic Y_2O_3, indicating a complex local structure due to a mixture of several species or the influence of the scatter O. The CN of Y-O shell (7.1) is larger than the nanoscale Y_2O_3 particles in the powder samples and the cubic Y_2O_3 (6), but consistent with the $YCrO_3$ model.

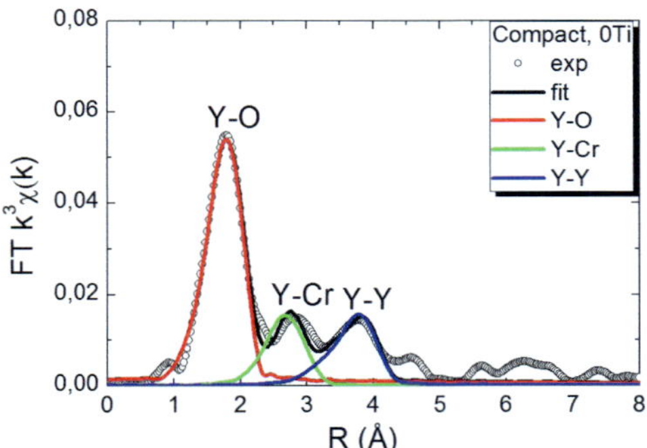

Figure 4.44: Fourier transform of EXAFS functions of 0% Ti compacted sample and data fit employing three coordinations: Y-O, Y-Cr and Y-Y. Phase shift was not corrected.

The compacted samples containing Ti have only two characteristic peaks in the FT (shown in Figure 4.43). The third peak is not distinct, making it difficult to specify which nanophases, Y_2O_3 or Y-Ti-O, exist. The alloy with 0.3% Ti seems to be consistent with Y_2O_3 with R_{Y-O}=2.27 Å and R_{Y-Y}=3.53 Å, but the high CN (8.5) of the first Y-O shell indicates that this is not a simple cubic Y_2O_3. As reported in the literature [92], a highly defected cubic phase Y_2O_3 with a space group Fm3m may form, the CN of the Y-O shell is 8, but the second shell of this phase with a R of 3.72 Å is much larger than that of 3.53 Å in the 0.3% Ti sample. Other compacted samples cannot be clearly identified as any of the single phases of the known Y-O and Y-Ti-O phases as mentioned above. It is likely that Y forms a mixture of various species, and none of them are dominant. As a spectroscopic technique, XAFS probes only an average over all species. The coordination peaks at high order shells are rather dispersed even for the compacted samples after HIP and annealing.

Table 4.11: The fitting results and experimental data for compacted ODS samples.

(a) Ti-free sample

Y K-edge	Y-O			Y-Cr			Y-Y		
	CN	R	σ^2	CN	R	σ^2	CN	R	σ^2
0 Ti	7.1	2.28	0.012	2.1	3.22	0.011	9.1	4.01	0.019

(b) Samples containing Ti

Y K-edge	Y-O			Y-Y		
	CN	R	σ^2	CN	R	σ^2
0.2Ti	6.5	2.27	0.0081	3.5	3.60	0.010
0.3Ti	8.5	2.27	0.0109	5.3	3.53	0.0146
0.4Ti	5.9	2.24	0.0093	2.2	3.62	0.010

4.4 Mechanical properties

4.4.1 Vickers hardness

Hardness is of direct interest as a measure of strength. The hardness tests have been performed on several polished metallographic cuts referred to as top, middle and bottom in Figure 3.4, respectively. The hardness results for various as-hipped alloys are given in Table 4.12 and each result is the average of 5 measurements. It is important to notice that the hardness distribution is macroscopically homogeneous with a maximum relative standard deviation (RSD) of 2.45 %.

Table 4.12: Vickers hardness for various hipped alloys.

	13.5Cr2W [kg/mm^2]				
	K12	K2	K4	K5	K6
Top	164	250	331	325	368
Middle	161	245	340	328	379
Bottom	169	242	340	322	383
Average	165	246	337	325	377
RSD %	2.45	1.65	1.54	0.92	2.06
	13.5Cr1.1W0.3Ti [kg/mm^2]		12Cr2W [kg/mm^2]		
	K13	K1	K15	K7	K8
Top	132	360	274	405	411
Middle	133	365	282	408	417
Bottom	133	360	274	411	413
Average	133	362	277	408	414
RSD %	0.44	0.80	1.67	0.74	0.74

The variation in hardness as a function of alloying elements is shown in Figure 4.45. After HIP, 12Cr base alloy K15 exhibits the higher hardness of 277 HV30 due to the presence of martensite in comparison to 13Cr base alloys K12 consisting of a fully ferritic structure. Low W concentration of 1.1% for K13 results in the lowest hardness. ODS steels show a remarkable increase in hardness by 47–172% and the highest hardness is found for 12Cr2W0.3Ti ODS alloy K8. For 13.5Cr2W steels, Ti plays an important role and the hardness increases generally with increasing the Ti concentration. The addition of 0.4% Ti leads to an increment from 246 HV30 (K2) to 377 HV30 (K6). The pre-alloyed 0.3% Ti in 13.5Cr1.1W ODS steel K1 compensates the loss of hardness by lower W concentration and thus leads to comparable hardness of 362 HV30.

Figure 4.46 show the influence of heat treatment on hardness for different ODS steels. As shown in Figure 4.46a, the hardness for base alloy K12 decreases by about 20% after annealing. This small amount of softening is attributed to the static recovery and the grain growth during annealing. A decrease in dislocation density and in grain boundary fraction is closely related to the annealing treatment and results in a drop of hardness. In contrast, a visible hardness enhancement is observed after annealing in all 13.5Cr2W ODS alloys. The non-Ti ODS alloy exhibits the largest increment in hardness of about 30%. The increase in hardness after annealing can be directly related to the elongated Cr carbides (Figure 4.7, Figure 4.13 and Figure 4.14) and thus enhanced solid solution. On the contrary, 12Cr2W steels in Figure 4.46b show an evident drop after tempering at 750 °C for 2 hours. The reduction in hardness is usually accompanied by an increase in ductility, thereby decreasing the brittleness. Precise control of time and temperature during the tempering process is critical to achieve the desired balance of strength and ductility.

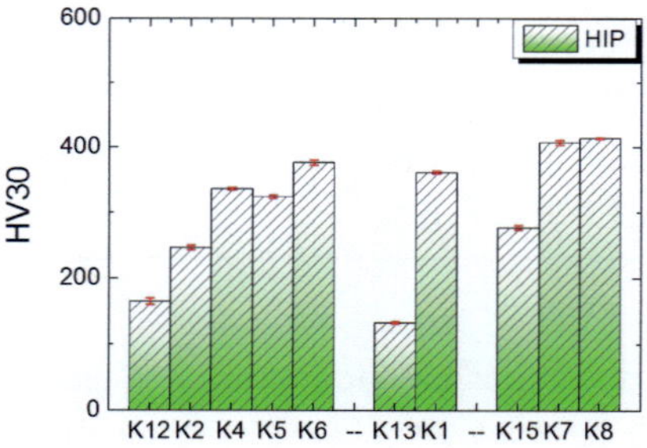

Figure 4.45: The comparison of the average Vickers hardness for various as-hipped alloys.

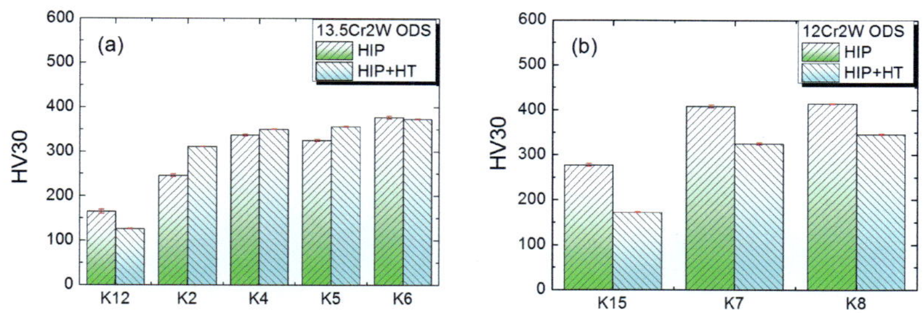

Figure 4.46: The influence of heat treatment on hardness for various ODS steels (a) 13.5Cr2W steels (b) 12Cr2W steels.

The comparison of hardness for 13.5Cr1.1W under different processing conditions is shown in Figure 4.47. The hardness for the base alloy K13 remains almost unchanged at different processing conditions. The ODS steel K1 decreases slightly in hardness after annealing. This reduction suggests that the softening by recovery and grain growth plays dominate role rather than the strengthening by the dissolution of Cr carbides. After TMP, the hardness is restored to the level of as-hipped state. The very close hardness values obtained from samples with different orientations indicate that TMP does not introduce evident anisotropy in hardness.

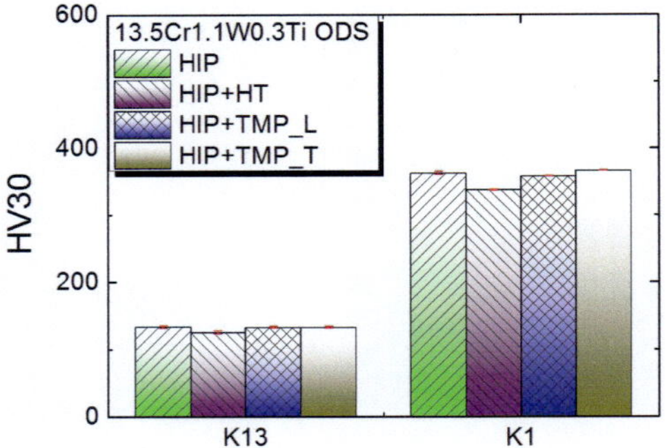

Figure 4.47: The comparison of hardness for 13.5Cr1.1W under different processing conditions. L: longitudinal direction; T: transverse direction.

4.4.2 Tensile properties

Figure 4.48 shows engineering stress-strain curves of 13.5Cr2W base steel and an ODS variant at different testing temperatures. The tensile strength of the base steel is 400 MPa and decreasing gradually as a function of increased temperature. The evident softening observed beyond 500 °C is characterized by the low tensile strength (approximately 60 MPa at 700 °C). The total elongation ranges from 30% to 70% and exhibits the converse temperature dependency compared to the strength. The addition of 0.3% Y_2O_3 and 0.4% Ti lead to an increase in strength from 400 MPa to 1100 MPa at room temperature. It is noted that the tensile strength for ODS steel at 600 °C is comparable to that for the base steel at room temperature. At 700 °C, ODS steel possess a tensile strength of 240 MPa, three times higher than base steel.

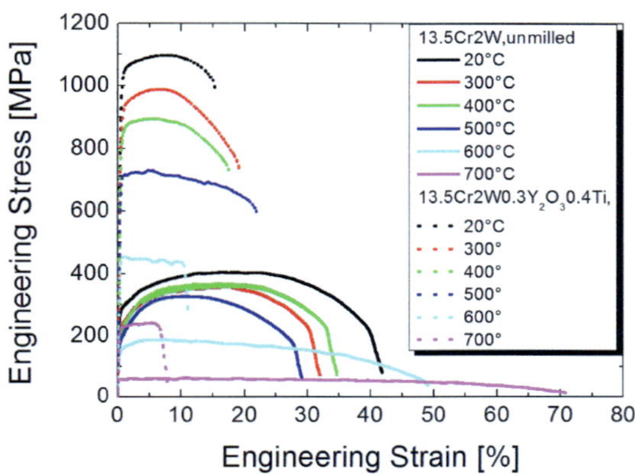

Figure 4.48: Comparison of engineering stress-strain curves of different steels at different testing temperatures.

The influence of Ti concentration on the tensile properties of ODS steels is shown in Figure 4.49. Tensile properties for all ODS alloys were remarkably enhanced and a more pronounced strengthening was observed for ODS ferritic steels in comparison with a 9% Cr ODS steel "FZK-ODS"over the test temperature range. Up to 500 °C the best tensile strength was observed for the ODS alloy with 0.3% Ti which was in accordance with particle size results. When increasing the test temperature further, Ti had a minor influence on tensile strength and all Ti-containing ODS alloys exhibited a comparable strength with the 9%Cr ODS steel. The ODS alloys with Ti show satisfactory strength up to 500 °C and a steep decrease in strength was clearly visible for these alloys above 500 °C. The yield strength varies in the similar manner to the tensile strength. However, 0.4Ti ODS alloy exhibits the highest yield strength among the whole temperature range (Figure 4.49b).

The enhancement in strength is usually accompanied by the degradation of the ductility, especially in terms of total elongation. As shown in Figure 4.49c, the ODS steel without Ti follows similar trend with EUROFER 97 and the maximum elongation of 37% is found at 600 °C. 0.4Ti ODS steel shows comparable total elongation up to 500 °C followed but a significantly drop is observed by increasing the temperature further. 0.3Ti ODS steel consists of the highest strength and the lowest elongation. The uniform elongation in Figure 4.49d is less sensitive to the testing temperature and usually below 10%.

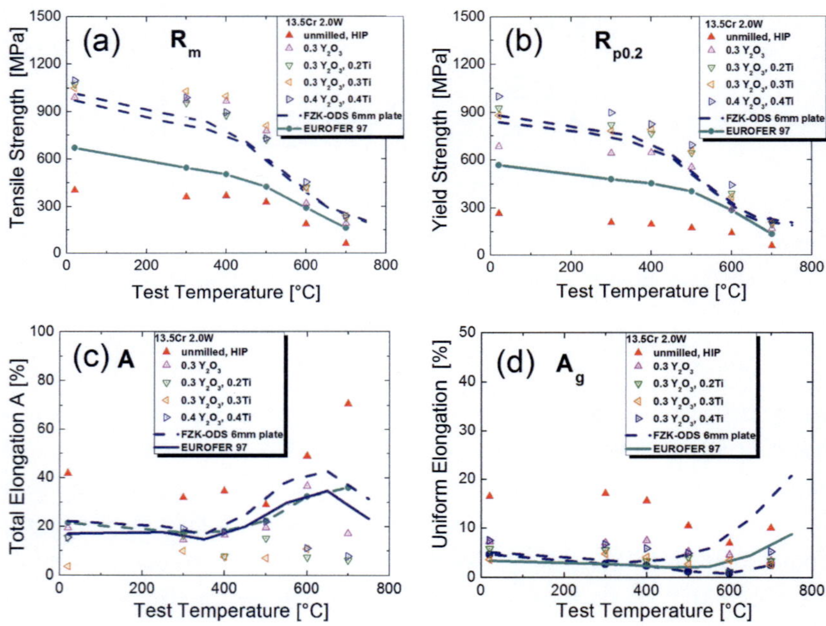

Figure 4.49: Comparison of the tensile properties for 13.5Cr2W ODS steels with different Ti concentrations as a function of testing temperatures (a) tensile strength; (b) yield strength; (c) total elongation; (d) uniform elongation.

In contrast to the highest hardness for 12Cr ODS steels, the tensile strength and yield strength for 12Cr ODS steels is approximately 10% lower than 13.5Cr2W ODS steel with 0.4 Ti. The addition of 0.3 Ti into 12Cr ODS steels has a more pronounced influence on the ductility rather than on the strength. The total elongation for 12Cr2W and 12Cr2W0.3Ti ODS steels is 10–20% and 6–9%, respectively. 13.5Cr2W0.4Ti ODS steel exhibit higher strength and better ductility compared to 12Cr2W ODS steels.

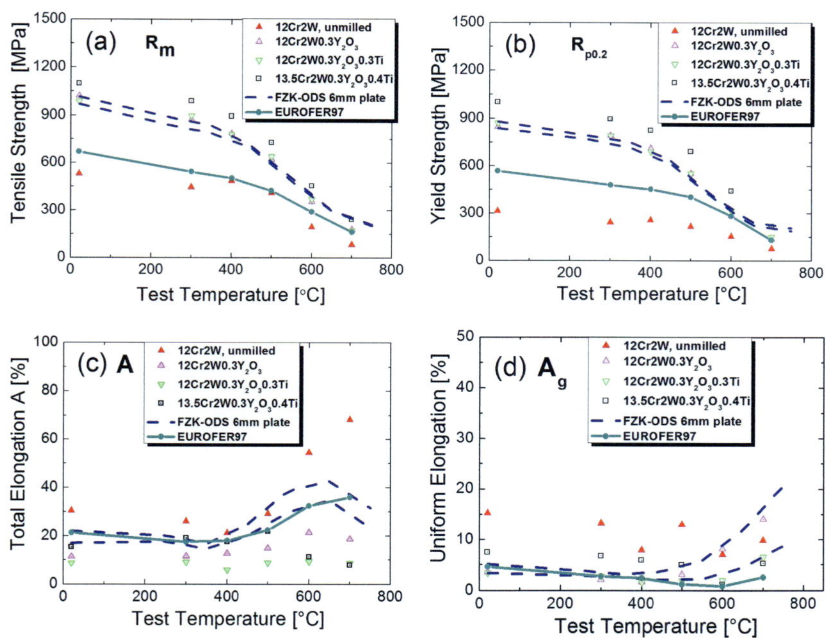

Figure 4.50: Comparison of the tensile properties for 12Cr2W ODS steels with 9Cr and 13.5Cr ODS steels as a function of testing temperatures (a) tensile strength; (b) yield strength; (c) total elongation; (d) uniform elongation.

The details of the tensile properties for 13.5Cr1.1W ODS steels are plotted in Figure 4.51. The tensile tests were performed on the specimens with both L-T and T-L orientations to evaluate the mechanical anisotropy induced by TMP. The tensile strength remains unchanged in both orientations in comparison to that for the corresponding HIP+HT ODS steels, whereas the yield strength seems slightly higher. More pronounced anisotropy is observed in total elongation and specimens with T-L orientation exhibits greater total elongation than other specimens. A maximum in elongation of 22% at 500 °C followed by a degradation of ductility at higher temperature is observed.

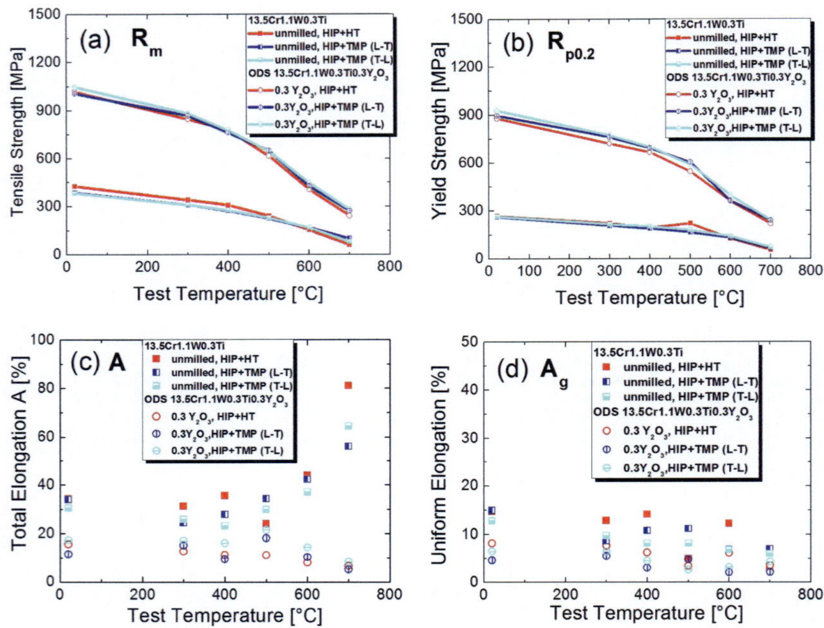

Figure 4.51: Comparison of the tensile properties for 13Cr1.1W ODS steels at different processing conditions as a function of testing temperatures (a) tensile strength; (b) yield strength; (c) total elongation; (d) uniform elongation.

Fracture morphology after tensile tests varies substantially depending on the material and processing conditions. Figure 4.52 shows how the fracture surface of 13.5Cr1.1W base alloy (K13) changes during processing. The HIP+HT specimen is broken in a full ductile manner with pronounced necking. A close look at the facture surface shows spherical dimples range from 1 µm to 50 µm in diameter, with some of them being relatively deep. The fracture surface for the L-T orientation specimen after HIP+TMP (Figure 4.52 c and d) clearly reveals a completely ductile fracture mode. The flat specimen found for the HIP+TMP steel indicates anisotropic deformation occurs during the tensile test. This is further confirmed by the elliptical dimples with aspect ratio of 2 to 4. This behavior is characteristic of ODS steels with elongated grain morphology introduced by hot extrusion or rolling. The crack propagation is more favorable along the main deformation direction than that along other directions.

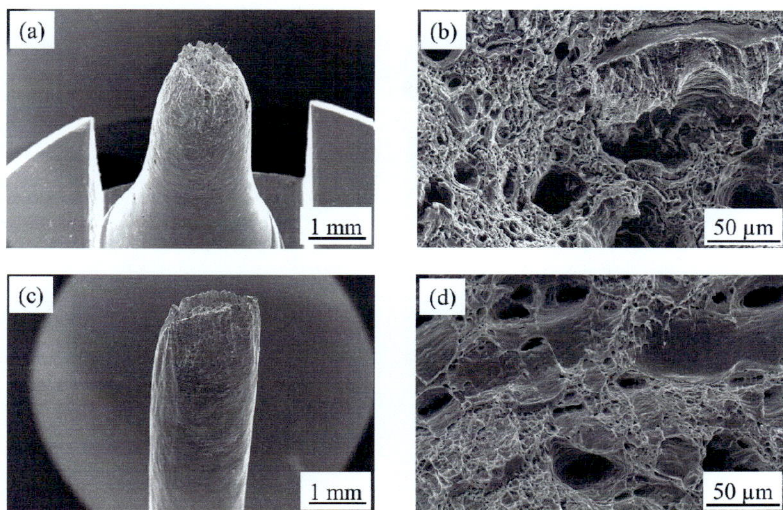

Figure 4.52: Fracture surface of the tensile specimens for 13.5Cr1.1W0.3Ti base steels deformed at room temperature (a) (b) HIP+HT; (c) (d) HIP+TMP, L-T orientation.

In case of the 13.5Cr1.1W ODS variant (K1) in Figure 4.53, the specimen fractures perpendicular to the specimen axis without observable necking. Furthermore the failure mode is predominately cleavage with some small area consisting of small and shallow dimples. This is in good accordance with the degradation in ductility for ODS steel.

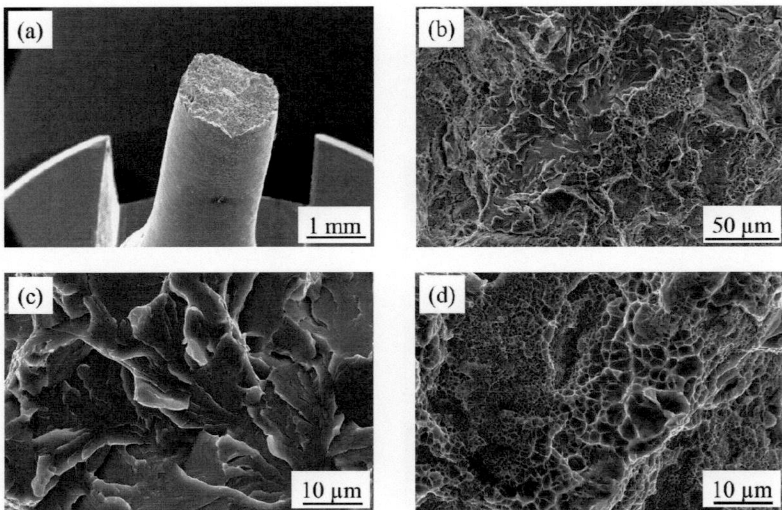

Figure 4.53: Fracture surface of the tensile specimens for 13.5Cr1.1W0.3Ti ODS steel after HIP+HT deformed at room temperature.

After TMP, the ODS steel exhibits slight deformation without evident necking. In contrast to the base material (Figure 4.52c), the fractured specimen remains cylindrical after the tensile test. Remarkable delaminations are found in the major part of the fracture surface of the L-T orientation specimen. Many local splits or laminations lie parallel to each other and perpendicular to the main rolling direction. The fracture surface displays alternating ductile and delaminated areas where the delamination mode plays a leading role. The initiation of the delamination in tensile test has not been sufficiently explained. It seems that elongated grain shape, certain texture characteristics, decohesion of grain boundaries, segregated impurity atoms and aligned particles and inclusions could lead to delamination either individually or cooperatively.

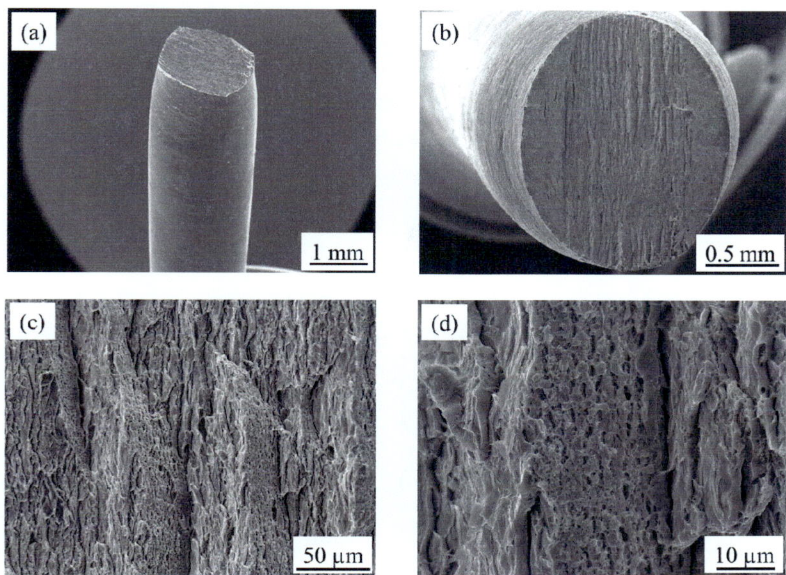

Figure 4.54: Fracture surface of the tensile specimens for 13.5Cr1.1W0.3Ti ODS steel after HIP+TMP deformed at room temperature.

4.4.3 Charpy impact properties

The effect of the chemical composition and processing history on impact properties is plotted in Figure 4.55 and Figure 4.56. In general, all the ODS specimens were broken into two pieces in the whole testing temperature range except Ti-free 13.5Cr2W ODS steel (K2). K2 only fractured at the temperature below room temperature suggesting superior impact properties in comparison to other ODS steels. The base steels remain unbroken at room temperature and elevated temperatures. K14 base alloy with L-T orientation does not fracture in the whole temperature range.

After HIP+HT, 13.5Cr2W base alloy exhibits the maximum impact energy of 8.4 J at room temperature and a DBTT of -40 to -20 °C. A decrease in USE by 25% and comparable DBTT are found for 13.5Cr2W ODS steel without Ti. The addition of Ti into ODS steels leads to a pronounced degradation of impact properties, particularly in terms of a much lower USE of about 1.4–1.8 J. In the case of 12Cr2W steel, the base alloy shows a reduction in USE of 2.7 J by a factor of 3 compared to 13.5Cr2W base alloy. The USE of 12Cr2W ODS steels lies around the similar level. This is due to the features of martensitic structure: martensite laths and high dislocation density. Therefore, 12Cr ODS steels are high strength materials, as confirmed by the Vickers hardness and tensile result, associated with inferior fracture toughness. Brittle fracture occurs over a wide range of temperatures.

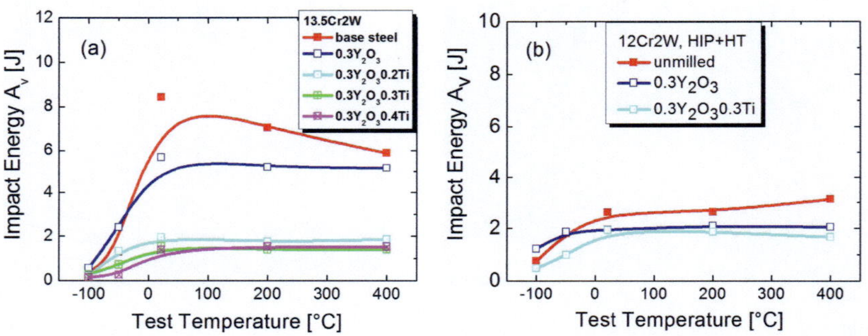

Figure 4.55: Temperature dependence of Charpy impact energy for ODS steels with different compositions.

Figure 4.56 presents the influence of TMP on the impact properties for 13.5Cr1.1W ODS steels. The samples in both L-T and T-L orientations were tested and compared. In contrast to EUROFER, TMP does not lead to pronounced improvement in impact properties for 13.5Cr1.1W base alloy. The L-T specimens show slightly higher USE than T-L specimens. For ODS variant, TMP results in an increase in USE by 30% for both L-T and T-L specimens. No obvious anisotropy is found due to the sample orientation.

The fracture surface of the different Charpy impact specimens have been analyzed by SEM and some typical results tested at room temperature have been shown in Figure 4.57, Figure 4.58, Figure 4.59, and Figure 4.60.

As shown in Figure 4.57, 13.5Cr2W base steel exhibits full ductile fracture at room temperature. The sample remains unbroken and pronounced shear lips are observed. The characteristic size of the dimples ranges from 1 to 10 µm, which is dependent on the size of the precipitates. The specimens exhibit similar features when increasing the testing temperature. Decreasing the testing temperature to -100 °C, the fracture surface represents a typical cleavage mode associated with cleavage facets and river patterns.

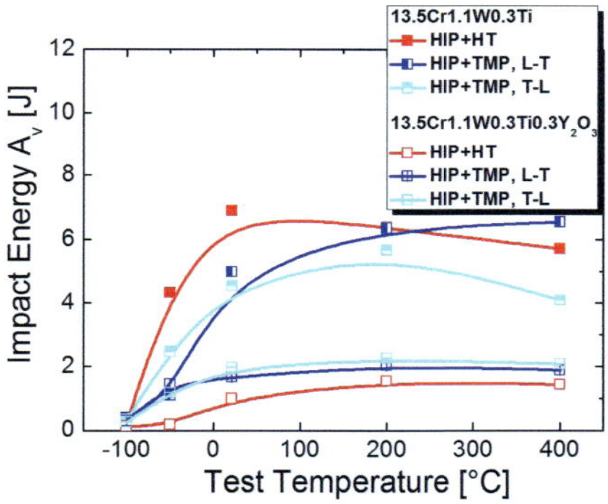

Figure 4.56: Temperature dependence of Charpy impact energy of 13.5Cr1.1W ODS steels at different processing conditions in comparison to the base steel.

In contrast to the base steel, the ODS steel K2 represents a mixed fracture mode at room temperature in Figure 4.58. The sample does not fracture and slight shear lips can be still found in Figure 4.58a. Nevertheless, the micro features are characteristic of a closely distributed mixture of cleavage facets and dimples (Figure 4.58b). The close distribution makes it difficult to quantify the ductile zone and the cleavage zone. Transgranular cleavage facets display a more or less pronounced river pattern. Tracing back the river pattern, the most probable cleavage initiation sites could be found. In most cases an area of a few facets must be considered as a cleavage initiation site. No small inclusion is found in the center of the initiating cleavage facet whereas numerous spherical precipitates are situated in the dimples. The spherical precipitates correspond to Cr oxides in Ti-free ODS steel K2 by EDX elemental mapping shown in Figure 4.33.

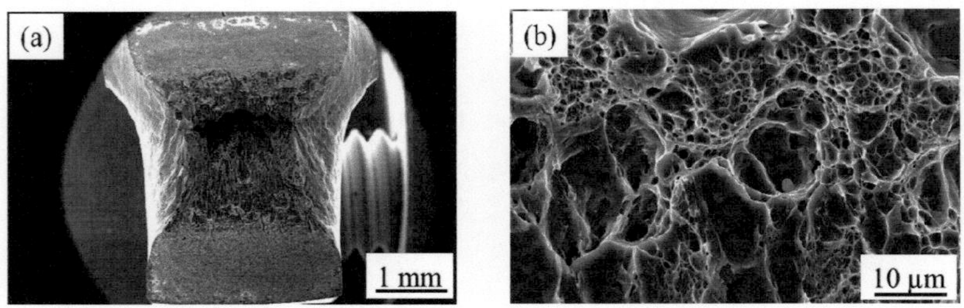

Figure 4.57: Fracture surface of 13.5Cr2W base steel K12 after tensile test at room temperature.

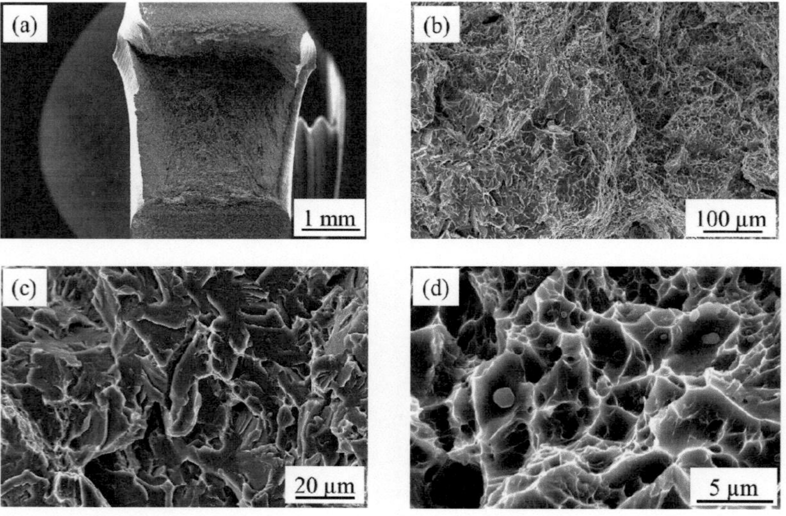

Figure 4.58: Fracture surface of 13.5Cr2W ODS steel K2 after tensile test at room temperature.

The addition of 0.4% Ti to ODS steel leads to a remarkable increase in strength and a pronounced degradation in USE. In relation to the fractography (Figure 4.59), the sample is completely broken after testing at room temperature without massive plastic deformation. The fracture surface is quite flat and transgranular cleavage facets play a dominate role.

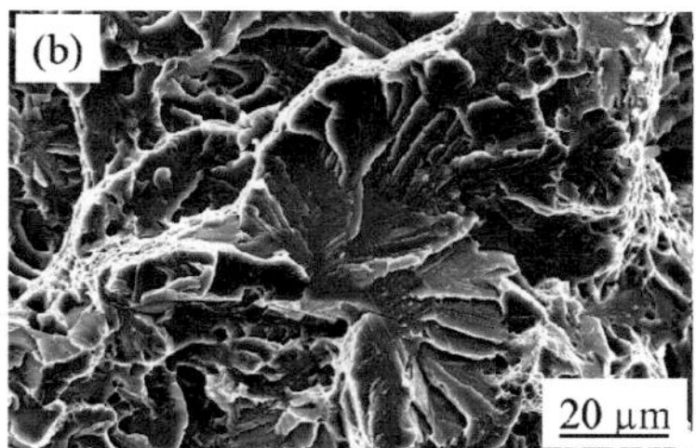

Figure 4.59: Fracture surface of 13.5Cr2W0.4Ti ODS steel K6 tested at room temperature.

TMP not only leads to an increase in USE by about 30% for 13.5Cr1.1W ODS steel, but also introduces pronounced delamination on the fracture surface (Figure 4.60). The most pronounced delamination is observed for the samples tested at room temperature in both L-T and T-L orientations. Multiple continuous deep splitting extends all the way to the back of the fracture surface. Some local splitting is also observed among these deep delaminations. Either increasing or decreasing the testing temperature, a significant decrease in both splitting number and length is observed. Frequently delamination is considered as a characteristic feature of the ultrafine elongated grain structures produced by heavy deformation. The influence of delamination on the impact property will be discussed in the next chapter.

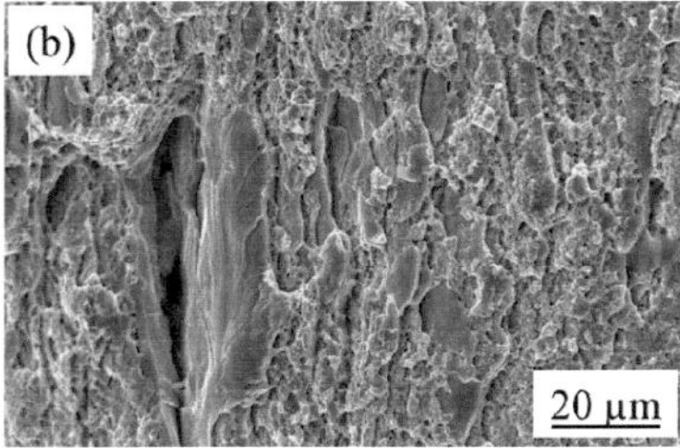

Figure 4.60: Fracture surface of 13.5Cr1.1W0.3Ti ODS steel after TMP tested at room temperature.

4.4.4 Fatigue properties

4.4.4.1 Surface roughness

The fatigue life until failure comprises two periods, the crack initiation period and the crack growth period. In many cases the crack initiation period covers a relative large percentage of fatigue life especially at low strain range. Fatigue in the crack initiation period is a surface phenomenon, which is very sensitive to various surface conditions, such as surface roughness. In order to reduce the data scatter and reveal the real fatigue life, surface roughness was measured by an optical sensor FRT MicroProf 150 at different polishing steps. Three different testing modes were employed to obtain reliable roughness results, as shown in Figure 4.61. A broad curved surface profile was measured in 3D mode (Figure 4.61b). Moreover, a flat surface profile with an area of 0.15×0.15 mm^2 and line profile were also performed at approximately the same height (Figure 4.16c).

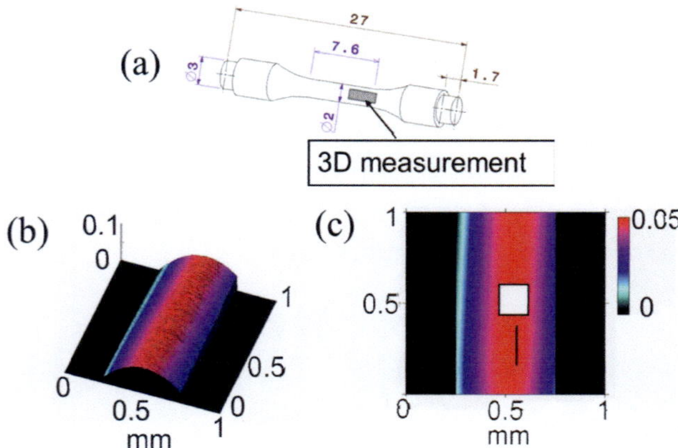

Figure 4.61: Schematic diagrams for different roughness testing modes by an optical sensor (a) 3D measurement; (b) corresponding roughness result for 3D measurement (c) 2D measurement and line scan.

Figure 4.62 shows the surface roughness for 13.5Cr1.1W ODS at different polishing processes. In order to improve the testing efficiency, 6 samples with different surface finish were used to evaluate the surface roughness at various polishing processes. The designation of the polishing process, for instance, K9-1-0, consists of three parts. K9 refers to 13.5Cr1.1W0.3Ti ODS steel after TMP. The first number in K9-1-0 only indicates that different samples are used for the roughness tests. The second number in K9-1-0 denotes the polishing process, for instance, 0 denotes as-received state, 1 to 4 denotes the corresponding mechanical processes (Table 3.7) and 5 refers to the final electropolishing process (Table 3.8). Four main surface roughness parameters were recorded and shown in Figure 4.62. The definition and the calculation method of these parameters can be found in Table 3.5. It is clearly visible that 3D mode represents the largest values of all these surface roughness parameters. R_z and R_{max} decrease when more polishing steps were applied. K9-4-3 shows the optimal R_z and R_{max} after three passes mechanical polishing. The further mechanical polishing and electropolishing processes lead to the surface roughening indicated by a raise of R_z and R_{max}. R_a and R_q are very stable during the whole preparation process. The surface roughness in 2D mode (Figure 4.62b) and line profile (Figure 4.62c) exhibits very similar trend during the polishing processes. In contrast to 3D mode, a dramatic drop in R_z and R_{max} is observed after electropolishing. Another difference is that 2D mode and line profile show smaller R_z and R_{max}. It is believed that the surface roughness should be improved when more polishing processes are employed. The unexpected optimal surface roughness observed by K9-4-3 could be attributed to the variations in surface state of as-received samples.

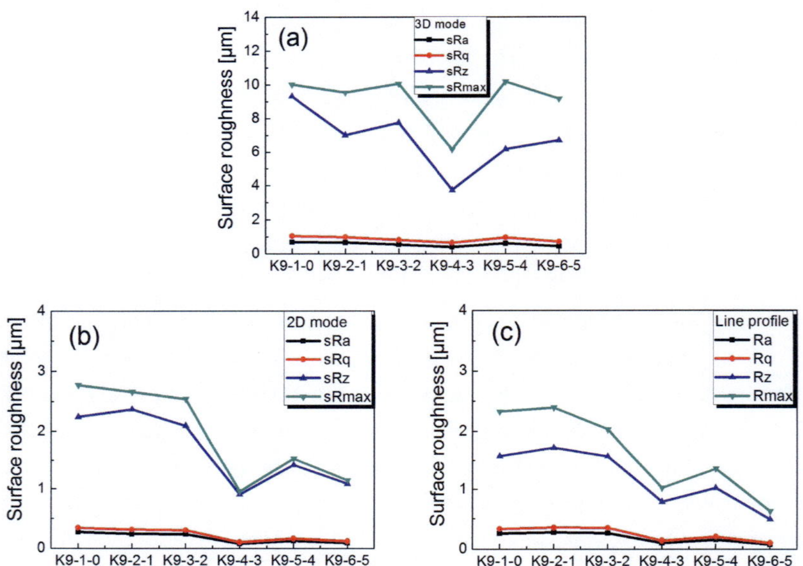

Figure 4.62: Surface roughness results for 13.5Cr1.1W ODS steels at different polishing steps (a) 3D surface roughness (b) 2D surface roughness and (c) line profile.

SEM micrographs of the polished samples are shown in Figure 4.63. The as-received sample K9-1-0 exhibits the worst surface with remarkable grooves both at low and high magnifications. The surface roughness is slightly improved after the first mechanical polishing process even through the grooves are still clearly visible at low magnification. From the second polishing process (K9-3-2) to the final electropolishing process (K9-6-5), smooth surface at low magnification is observed. The polishing processes reduce the grooves significantly. After electropolishing, the sample surface is free of noticeable grooves or cracks even at high magnification, whereas very few voids are still visible.

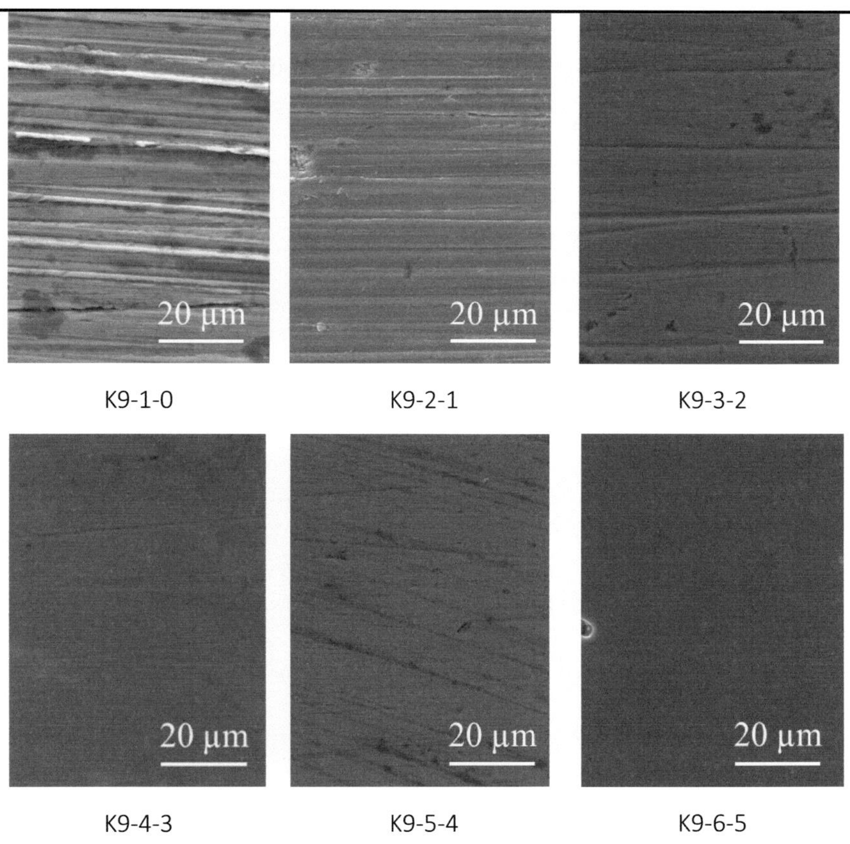

Figure 4.63: SEM micrographs of the samples under different polishing conditions. The details of the polishing processes is available in Table 3.7 and Table 3.8.

4.4.4.2 Fatigue results

Figure 4.64 shows the total, plastic and elastic strains as a function of number of cycles to failure for 13.5Cr1.1W0.3Ti ODS steel K9. The total strain is composed of the elastic deformation and plastic deformation. The plastic deformation reaches ~ 0.3% at total strain of 0.9% and 0.04% at total strain of 0.54%. Simple linear relationship could be obtained for both total strain and elastic strain in the whole strain range. The slopes for total strain and elastic strain are -0.10 and -0.06, respectively. The plastic deformation seems to follow a linear Coffin-Manson relationship with a slope of -0.40.

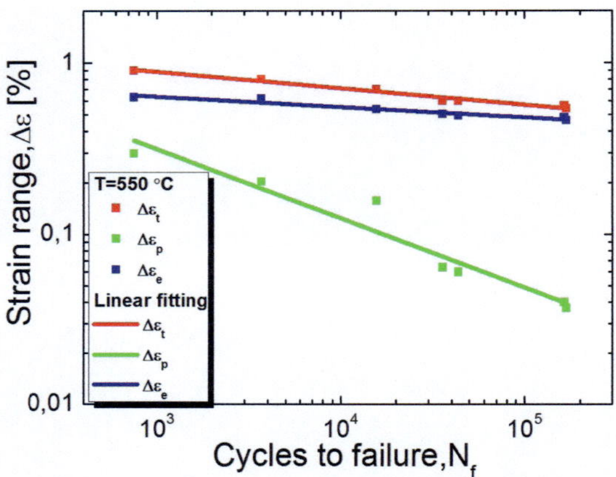

Figure 4.64: Total, plastic and elastic strains versus number of cycles to failure at 550 °C under strain-controlled fatigue tests for 13.5Cr1.1W0.3Ti ODS steel K9.

Typical S-N curves at different total strain range are shown in Figure 4.65. The stress amplitude in the whole lifetime ranging from 10^3 to 1.7×10^5 cycles is constant irrespective of the strain range. Neither cyclic softening nor cyclic hardening is observed indicating the microstructural stability of ODS steel during cyclic deformation at 550 °C.

The stable stress during the entire fatigue test is further confirmed by the hysteresis loops at various fraction of fatigue life, as shown in Figure 4.66. Four specific fractions between the very beginning and $0.95N_f$ of the lifetime are chosen and the hysteresis loops almost overlap with each other. At lower strain region (up to $\Delta\varepsilon_t \leqslant 0.6\%$), only very slight amount of plastic deformation is present. When increasing the strain further, a profound increase in plastic strain is observed and the ratio of plastic deformation over elastic deformation at $\Delta\varepsilon_t = 0.9\%$ reaches 30%.

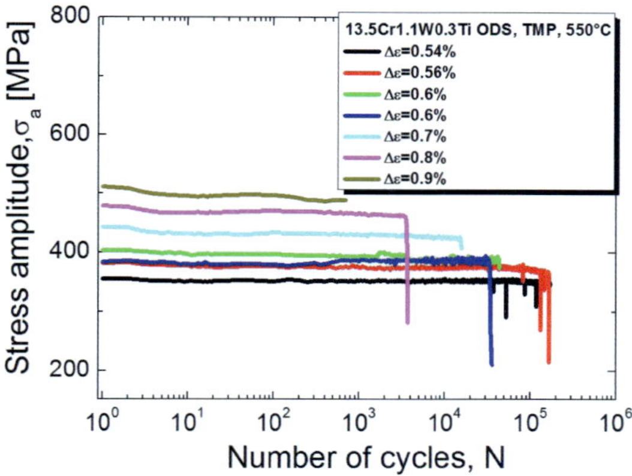

Figure 4.65: Stress amplitude as a function of number of cycles under strain-controlled fatigue tests at 550 °C for 13.5Cr1.1W0.3Ti ODS steel K9.

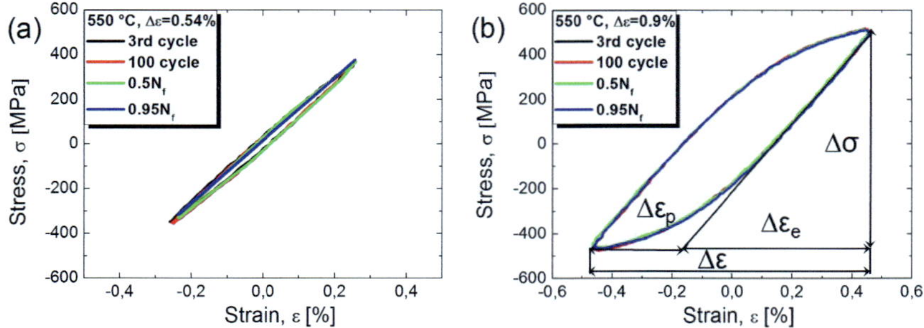

Figure 4.66: Hysteresis loops of 13.5Cr1.1W ODS steels K9 under various strain ranges (a) $\Delta\varepsilon$=0.54% (b) $\Delta\varepsilon$=0.9%.

Figure 4.67 shows total, plastic and elastic strains as a function of number of cycles under various strain ranges. The ratio of the plastic and elastic strain remains unchanged in the whole lifetime. This is in good agreement with the overlapping of hysteresis loops at various stages of the fatigue test in Figure 4.66. Both the constant stress and strain during the test suggests that ODS is characteristic of very stable structure. When increasing the total strain, a greater increment in plastic deformation is found.

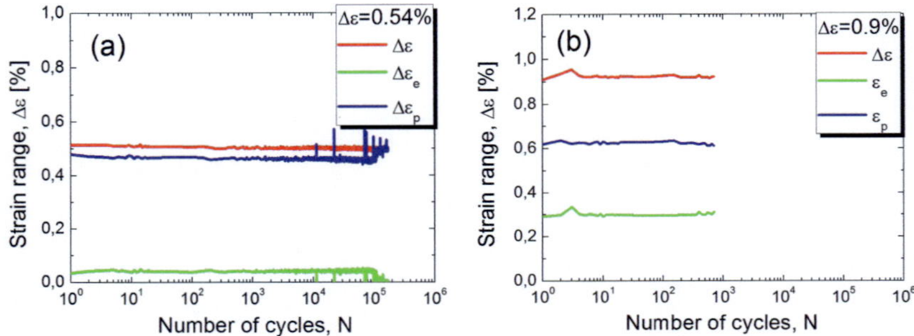

Figure 4.67: Total, plastic and elastic strains versus number of cycles at 550 °C under various strain ranges (a) Δε=0.54% (b) Δε=0.9% for 13.5Cr1.1W0.3Ti ODS steel K9.

4.4.4.3 Microstructure analysis

Figure 4.68 shows the microstructure and HAADF results of 13.5Cr1.1W ODS steel after low cycle fatigue (LCF) test at Δε=0.9%. LCF at 550 °C does not change the elongated grain morphology induced by TMP significantly. Numerous rod-shaped or round precipitates with different contrasts are visible along the grain boundaries as well as within the grains. No pronounced chemical segregation is introduced by LCF test. The EDX spectra of the bright rod-shaped precipitate P1 and the black round precipitate P2 are recorded in Figure 4.68c and d, respectively. Both precipitates are Ti-enriched precipitates. The difference in contrast can be attributed to the different ratio of (Fe+Cr)/Ti. The bright rod-shaped precipitate P1 exhibits a high concentration of matrix (Fe and Cr) indicating that it is embedded in the matrix. On the contrary, only very weak signal of Fe and Cr is observed for the black round Ti-precipitate P2. This fact suggests that this precipitate is on the surface of the foil. The analytical investigation on other bright round precipitates revealing similar (Fe+Cr)/Ti ratio with Figure 4.68c, is not shown here.

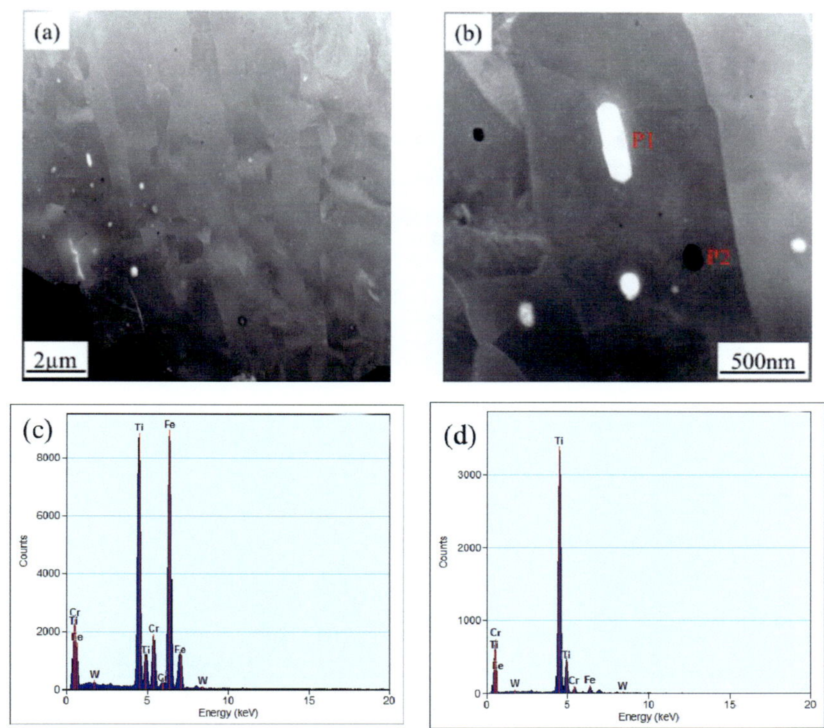

Figure 4.68: HAAADF images of general view (a) and coarse precipitates (b) in 13.5Cr1.1W0.3Ti ODS steel K9 at LCF test at Δε=0.9%. The EDX spectra for the two precipitates marked with P1 and P2 are shown in (c) and (d) respectively.

There is no doubt that the high dislocation density in ODS ferritic steel and its evolution is very important to understand the constant stress amplitude during the whole LCF test process irrespective of the strain range. Typical values for dislocation densities in pure metals and alloys are 10^{12} m^{-2} (undeformed) and 10^{16} m^{-2} (after cold working). It can be seen from Figure 4.69 that the dislocation density is very high. However, dislocation density can vary, not only between different grains but also within one micro grain. Here the dislocation density for one material state is presented by the mean value of 3 individual measurements. The dislocation densities and RSD are summarized in Table 4.13. Prior to LCF test, the dislocation density is approximately 7.75×10^{14} with a relative standard deviation (RSD) of 47%. After LCF test at $\Delta\varepsilon$=0.9%, the dislocation density maintain the same order of magnitude with lower RSD of 25%. This suggests that dislocation density is more homogenous after LCF. The ODS ferritic steel exhibits an intermediate value of dislocation density between the pure metals and the alloy after cold working. The constant dislocation density before and after LCF test can explain the stable stress amplitude.

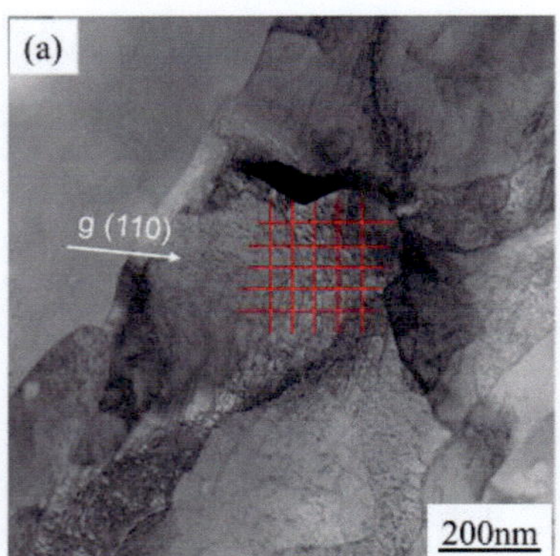

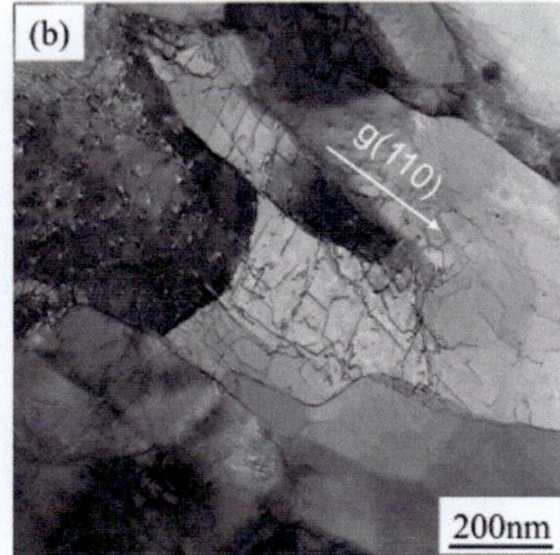

Figure 4.69: The dislocation distribution in 13.5Cr1.1W0.3Ti ODS steel K9 before (a) and after LCF test at $\Delta\varepsilon$=0.9% (b).

Table 4.13: Dislocation densities and RSD for various specimen locations and material states.

Number	Dislocation density before LCF [m^{-2}]	Dislocation density after LCF [m^{-2}]
1	1.15×10^{15}	7.0×10^{14}
2	4.21×10^{14}	4.77×10^{14}
3	7.56×10^{14}	8.05×10^{14}
Mean dislocation density	7.75×10^{14}	6.6×10^{14}
Relative standard deviation (RSD)	47%	25%

In ODS steels, free dislocations represent both carriers of plastic deformation and obstacles to plastic deformation. And both characteristics may govern the behavior of ODS steels depending on the stage of the fatigue process. As shown in Figure 4.70, a pronounced particle-dislocation interaction is present after LCF test. Figure 4.70a shows that the dislocations are pinned by the spherical particles with a size of 5–10 nm in diameter. The contrast from the dislocation makes it impossible to distinguish the particles smaller than 5 nm involving in the dislocation pinning. No particle deformation is found in ODS ferritic steels after LCF test by TEM. This may be accounted for the great strength of Y_2O_3 and Y-Ti-O particles. Despite the absence of Orowan loops, the typical particle-dislocation characteristics at elevated temperature, departure side pinning, is clearly visible. In contrast to the Orowan model, at elevated temperatures, the dislocations are attached to the particles (Ti oxides), some with very sharp contact angles. The multiple precipitate-dislocation interaction and dislocation-dislocation interaction are presented in Figure 4.70b.

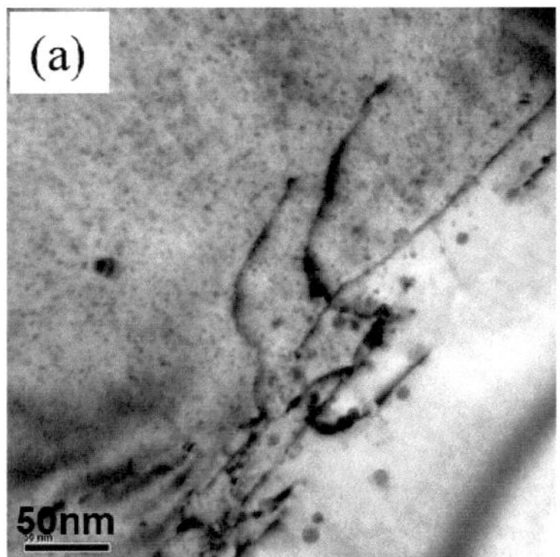

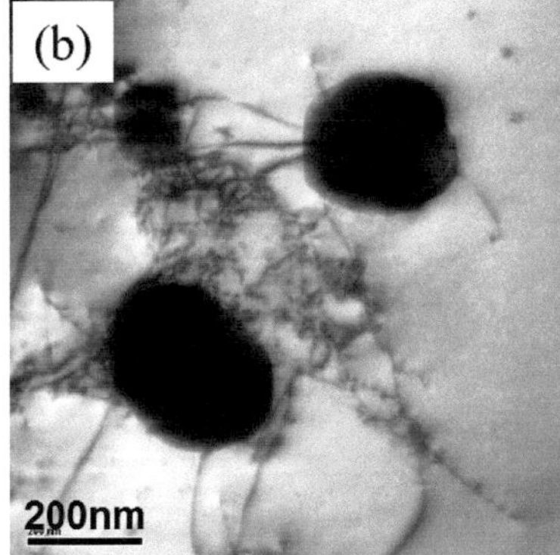

Figure 4.70: Particle-dislocation interaction in 13.5Cr1.1W0.3Ti ODS steel K9 after LCF test at $\Delta\varepsilon$=0.9%. The ultrafine dispersoids in (a) are Y-Ti-O particles and the coarse precipitates in (b) are Ti oxides.

4.4.4.4 Fractography

There are two modes for the tester to determine the end of the fatigue. In case of high strain range, the fatigue life is relatively short, generally less than 10^5 cycles. Therefore the fatigue test under high strain range will last until the sample is broken. The fatigue life at low strain range can reach up to 10^6–10^7 cycles. Under these circumstances, the main criterion for the end of the fatigue test is the ratio of the maximum stress for the last cycle to that for the first cycle of the test. When the ratio is approximately 30%, the test will stop.

Among all the LCF tests, only the fatigue life for the sample under the lowest strain range 0.54% is determined by the ratio of the maximum stress. And the sample remains unbroken after unloading, as shown in Figure 4.71a.

The crack origin process can be deduced from the morphology of the lateral surface for the sample at $\Delta\varepsilon$=0.54% in Figure 4.71. From the macroscopic view in Figure 4.71b, the surface becomes rougher in comparison to the sample after electro polishing (Figure 4.63). The dark dots and rods have the same chemical composition with the matrix and they may be resulted from the heterogeneous deformation. The closer view of the lateral surface in Figure 4.71c reveals numerous microcracks almost parallel to the loading direction. The microcracks are believed to initiate at coarse precipitates or micropores indicated by the arrows. The microcracks propagate preferably by the coalescence of these micropores. The aggregate boundaries are favorable sites for crack initiation and crack propagation. The microcracks in Figure 4.71c show a perfect correlation with the elongated grain structure.

Figure 4.72 presents the lateral surface of the sample tested at $\Delta\varepsilon$=0.9%. From the macroscopic view, a V shaped fracture surface is observed with an angle of 30–40° to the loading direction. Multiple crack initiation sites are found on the surface, with coarse Ti precipitate (indicated by the arrow in Figure 4.72b) or without the presence of precipitates (Figure 4.72c). In the vicinity of the crack initiation site with a precipitate, two deep main cracks are clearly visible with remarkable intrusions and extrusions (Figure 4.72d). In contrast, only numerous crack segments with micropores are found close to the crack initiation site without precipitates and the surface seems slightly smooth. The intrusions and extrusions are still visible but not that distinct.

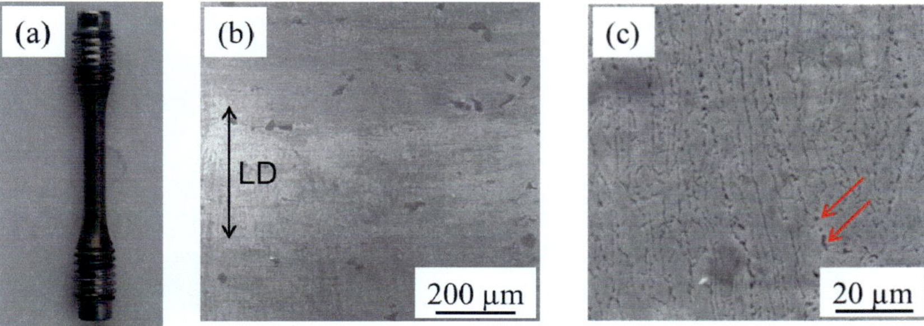

Figure 4.71: Lateral surface of the sample 13.5Cr1.1W0.3Ti ODS steel K9 tested at Δε=0.54% (a) overview; (b) macroscopic view and (c) microscopic view.

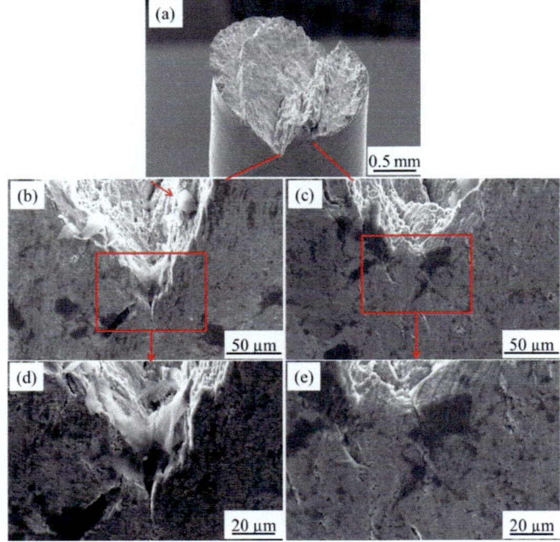

Figure 4.72: Lateral surface of the sample 13.5Cr1.1W0.3Ti ODS steel K9 tested at Δε=0.9% (a) macroscopic view, (b) (c) two crack initiation sites, (d) (e) microscopic view of (b) and (c) respectively.

Figure 4.73 shows the fracture surface of 13.5Cr1.1W ODS steel tested at $\Delta\varepsilon$=0.9%. Two crack nucleation sites are indicated by the arrows in Figure 4.73a. Several deep cracks with a length of 0.5–1.0 mm propagate from the surface towards the interior of the sample. The local cracks are present near the main crack. The typical feature, striations are absent in the crack propagation region. The final fracture region (Figure 4.73c) away from the deep cracks exhibits the same tear trend and dimples as the tensile fracture surface at 600 °C (Figure 4.73d).

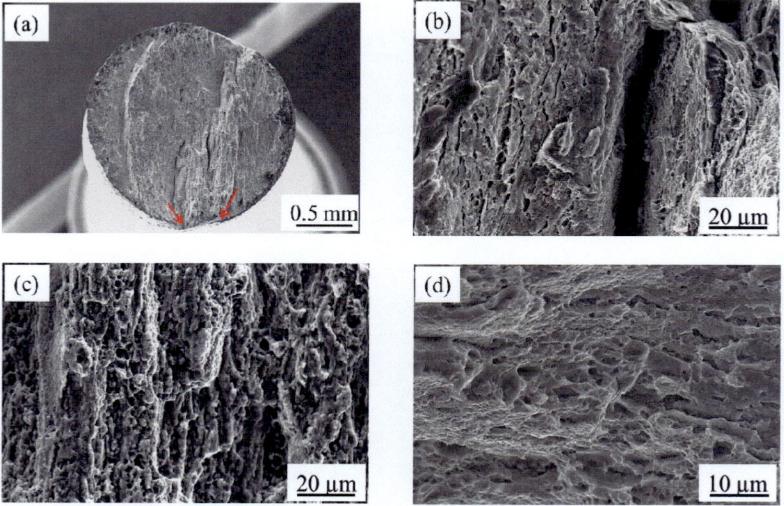

Figure 4.73: Fracture surface of 13.5Cr1.1W0.3Ti ODS steel after LCF test at 550 °C at $\Delta\varepsilon$=0.9% (a) macroscopic view; (b) deep crack inside the sample; (c) dimples in final fracture area and (d) tensile fracture surface at 600 °C.

5 Discussion

5.1 Influence of chemical composition

5.1.1 Influence of chemical composition on microstructure

The primary goal of this work is to qualify high-performance alloy for structural application of blanket and divertor in fusion system. Simplicity of the alloy composition facilitates the processing and controlling during the fabrication procedure, and enables also more straightforward understanding of the role of different alloying elements.

In this work, pure ferritic steel, with bcc crystal structure, must be preserved to extend the window temperature of ODS steel (>700 °C). The γ-loop in the Fe-Cr equilibrium diagram is closed if the chromium content exceeds about 12% at all temperature (Figure 5.1). Cr in an iron alloy provides strength, oxide resistance and acts as a ferrite stabilizer. However, higher Cr contents promote the formation of σ-FeCr phase as well as Cr rich α' phase, which are responsible for brittleness, inferior creep resistance and corrosion sensitivity. Therefore, 12% Cr is often chosen to ensure a fully ferritic microstructure with minimal Cr content.

Cr is medium carbide former. Figure 5.2 shows that different types of Cr carbides exist in Fe-Cr-C system as a function of carbon content. Carbide modification $M_{23}C_6$ is mostly located in the low carbon region (C% < 1%) below 850 °C. Other carbides, for instance, M_7C_3 and M_3C, are found with the increase of carbon content. "M" stands for random occupancy of the Fe lattice sites with Fe and other metal atoms substituting for Fe in the solid solution. As shown in Figure 4.7 and Figure 4.10, the grain boundaries for 13.5Cr2W ODS steels (K2, K4, K5, K6) and 13.5Cr1.1W0.3Ti ODS steel (K1) are decorated with elongated Cr-rich carbides after HIP. Although high resolution TEM has not been performed on Cr carbides in this work, the morphology and the analytical results by means of EDX and EELS are in good agreement with $(Fe,Cr)_{23}C_6$ type observed in "as-HIPped" EUROFER ODS steel [68]. As indicated in Figure 5.2,

$M_{23}C_6$ forms below 850 °C, which is sufficiently lower than the HIP temperature (1150 °C). The presence of the Cr carbides in the as-HIPped ODS ferritic steels can be accounted for a slow cooling rate after HIP and thus for an accelerated diffusion rate. Moreover, the ultra-fine grain structure for ODS steels enables fast diffusion along grain boundaries. This is further supported by the segregation of Cr carbides and Cr-depleted areas at grain boundaries. Annealing treatment at 1050°C with a much faster cooling rate is very effective to dissolve Cr carbides, as shown in Figure 4.13 and Figure 4.14.

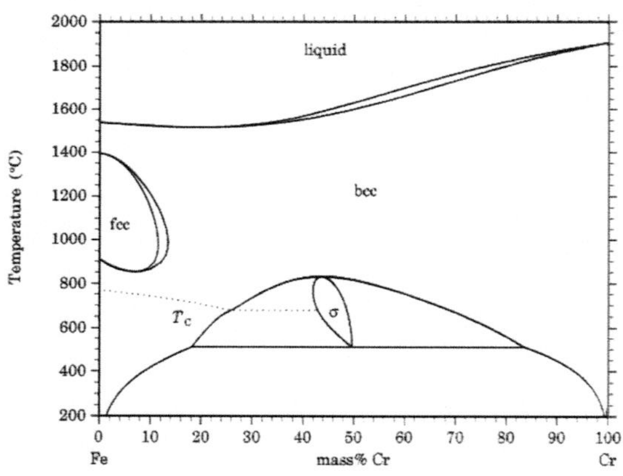

Figure 5.1: The constitution of Fe-Cr alloys [93].

The Fe-Cr-W phase diagram with 2% W, suggests that 12–14% Cr ODS ferritic steels may consist of α–Fe rich bcc' matrix and several intermetallic phases such as μ Fe_7W_6 and Fe_2W Laves phase (C14), as shown in Figure 5.3. The Cr rich bcc'' phase tends to coexist with the Fe rich bcc' phase by increasing Cr content. Addition of W stabilizes the σ-phase from the Fe-Cr side into the ternary system. The χ-phase with α-Mn structure is present at elevated temperature between 570 °C and 780 °C. Laves phase distributed preferably either on the grain boundaries [94]or was located close to $M_{23}C_6$ carbides [95] in conventional 9%Cr steels. Narita et al. [96] revealed that appropriate W concentration of 2% improves high temperature strength without introducing brittle Laves phase and/or μ phase for 9%Cr ODS steels. EDX elemental mapping in Section 4.2.2 shows that Fe_2W Laves phase is absent in 12–13.5%Cr ODS steels.

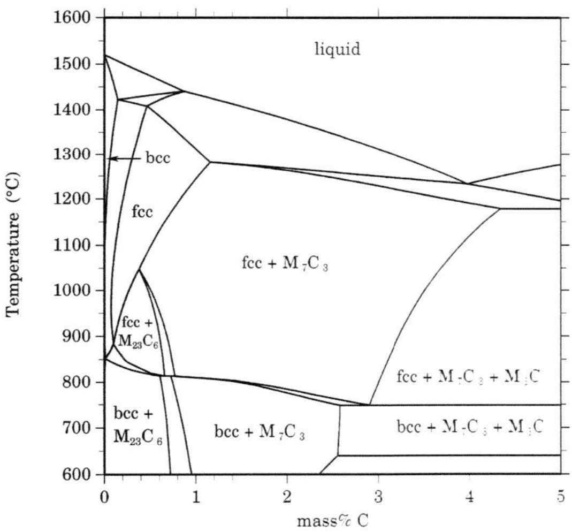

Figure 5.2: Effect of carbon on the constitution of Fe-Cr-C alloys containing 13% Cr [93].

The ternary phase diagram Fe-Cr-W in Figure 5.4 provides a comprehensive view of the main alloying elements for the ODS steels. The phase diagram consists of the broad bcc phase whereas the fcc phase exists only in a restricted range of temperature and composition at the Fe corner. All the ODS steels in Table 3.2 are located in the bcc phase according to the chemical composition. However, the austenite-stabilizing elements (C, N, Ni, Mn, Cu, Co) which extend the γ-phase field must be taken into account. In addition, the phase diagrams only represent the stability of the phases under equilibrium condition. Other metastable phases like martensite are not included in the phase diagram.

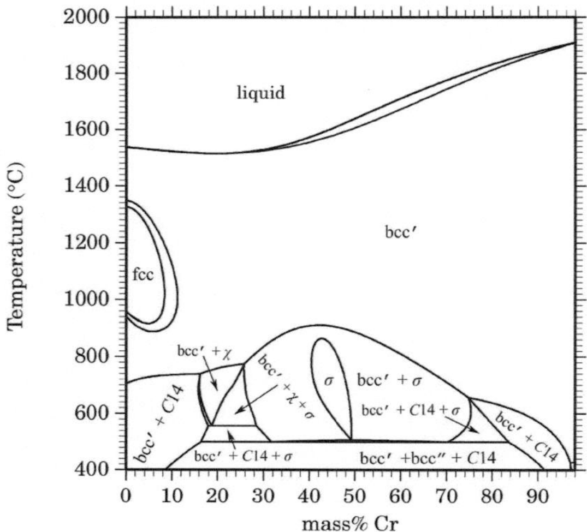

Figure 5.3: Effect of chromium on the constitution of Fe-Cr-W alloying containing 2% W [93].

An attempt was made to predict the constitution of the steels at room temperature with a Schaeffler-Schneider diagram using the following Ni and Cr equivalents of the alloying elements [54]:

$$Ni_{equiv} = Ni + Co + 0.5(Mn) + 0.3(Cu) + 30(C) + 25(N) \tag{5.1}$$

$$Cr_{equiv} = Cr + 2(Si) + 1.5(Mo) + 5(V) + 1.75(Nb) + 0.75(W) + 1.5(Ti) + 5.5(Al) \tag{5.2}$$

The Ni and Cr equivalents are calculated according to Equation (5.1) and (5.2) based on the measured chemical composition. The Schaeffler-Schneider diagram for various alloy systems is shown in Figure 5.5. 13.5Cr1.1W alloy system is predicted to be fully ferritic while other alloys, for instance 12Cr2W alloys and 13.5Cr2W alloys, tend to have both martensite and δ-ferrite. Ferritic-martensitic structure of 12Cr2W alloys is consistent with the prediction due to the unexpected high Ni content (1.6%). Moreover, the formation of δ-ferrite was successfully observed in 9%Cr ODS steel powder using EPMA observation and in-situ XRD/SAXS techniques. The residual α-ferrite phase coexists with austenite phase in the fully austenite area (1150 °C) due to incomplete transformation [97]. For ODS steels, Y-Ti complex oxide particles precipitate at elevated temperature ≥ 700°C. The dilatometric measurements show that α→γ transformation occurs over 750 °C (Table 4.3). The retention of residual α-ferrite phase can be attributed to the formation of Y-Ti complex oxide particles. These particles block the motion of the α-γ interface, thereby suppressing the complete α→γ transformation [98]. The evident variations in M_s and M_f for three 12Cr2W steels (K15, K7, K8), as shown in Table 4.3, indicate that the addition of Y_2O_3 and Ti have profound influence on the kinetics of martensitic transformation.

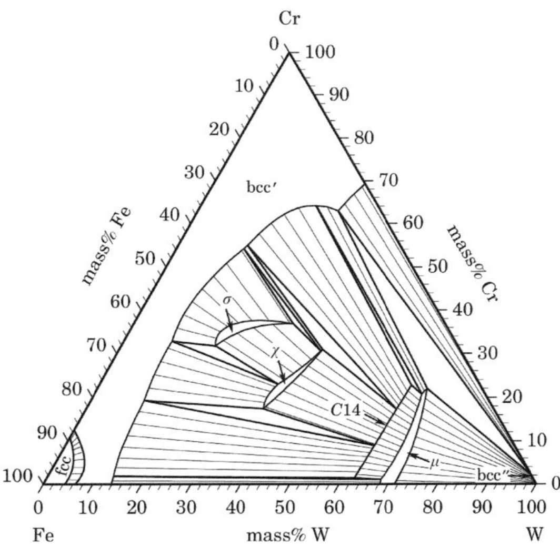

Figure 5.4: Isothermal section of Fe-Cr-W system at 1100 °C [93].

The microstructural observation reveals fully ferritic structure in 13.5Cr1.1W alloys. It is evident from Equation (5.2) that both carbon and nitrogen have a significant impact on the stability of ferrite. Although similar Cr equivalent is observed for 13.5Cr2W and 13.5Cr1.1W0.3Ti alloys, higher C and N contents result in greater Ni equivalent and therefore 13.5Cr2W alloys lies in the dual-phase region. However, 13.5Cr2W alloys exhibits fully ferritic structure with large amount of Cr carbides along the grain boundaries (Figure 4.7). This inconsistence can be explained by the formation of carbides. The ferrite-forming element Cr is effective to remove the austenite former C and/or N from solution to insoluble carbides and nitrides.

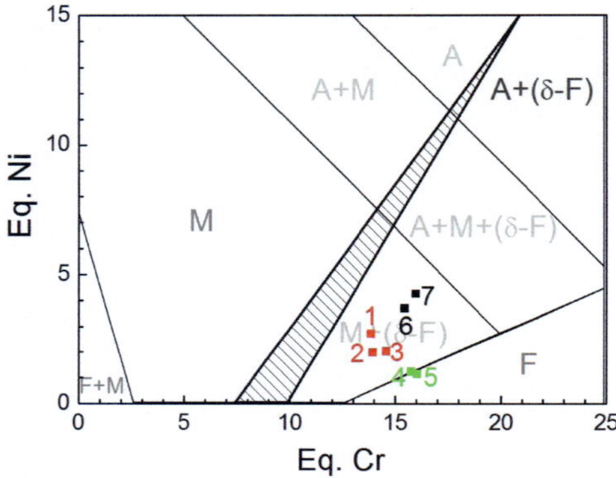

Figure 5.5: Schaeffler-Schneider diagram for different alloy systems (1) 12Cr2W system: 1(K7), 2 (K15) and 3(K8); (2) 13.5Cr1.1W0.3Ti system: 4(K1) and 5(K13) (3) 13.5Cr2W system: 6(K2) and 7(K5).

5.1.2 Influence of chemical composition on mechanical properties

The presence of a ferritic/martensitic structure in 12Cr ODS steel leads to the highest hardness (Table 4.12) after HIP. During the martensitic transformation, the fcc austenite transforms to a highly strained bcc form of ferrite which is supersaturated with carbon. With a comparable content of carbon in all hipped ODS steels, large numbers of dislocation caused by shear deformation is considered as a primary strengthening mechanism for 12Cr ODS steels. The tempering treatment at 750 °C for 2 hours leads to a remarkable decrease in hardness for 12Cr ODS steels (Figure 4.46b). This is also reflected by a reduced tensile strength and yield strength (Figure 4.50). The softening can be explained by the recovery during tempering. In 12Cr ODS steels, the dislocations are trapped at lath boundaries (Figure 4.12). The kinetics of recovery is associated with the spreading of trapped lattice dislocations into grain boundaries. The decrease in dislocation density leads to no significant accumulation of dislocations inside the grains and, consequently, to less work hardening.

It has been reported [34] that the addition of low-activation W provides uniform solid solution and thus improves the tensile strength and creep resistance. The tensile strength as a function of testing temperature has been compared for several typical experimental and commercial ODS steels, as shown in Figure 5.6. A significant variation in tensile strength of different steels occurred for tests at room temperature to 1000 °C. Among all these alloys, 14YWT exhibit the highest tensile strength over the whole temperature range whereas PM2000 is the weakest one. Obvious increase in tensile strength between 300 °C to 500 °C is observed for 13.5Cr2W0.3Ti ODS steel K5 compared to 13.5Cr1.1W0.3Ti ODS steel K1 up to 600 °C. EUROFER ODS shows almost the same trend with K1. Both 12YWT and MA957 are stronger than our experimental steels, K5 and K1. Although the tensile strength is strongly dependent on numerous factors, for instance, fabrication history, sample orientation, it can be still seen that higher content of W makes considerable contribution to tensile strength due to solid solution. The wide ranges of creep strength as well as the strength advantages of MA957 and 12YWT have been reviewed with comparison to EUROFER 97 [20].

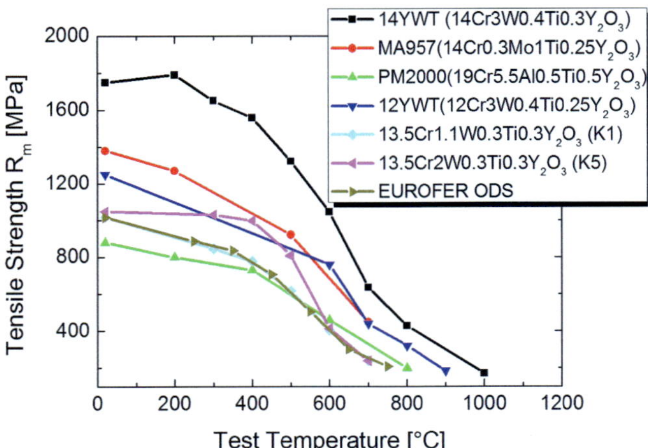

Figure 5.6: A comparison of ultimate tensile strength as a function of testing temperature for the experimental and commercial ODS steels [16, 34, 99].

It has been revealed that the addition of Ti has a significant influence on the size and the type of nanoscale oxide particles. This applies also to the current case. As shown in Figure 4.49, the best tensile strength was observed for the ODS alloy with 0.3% Ti up to 500 °C. This is in good agreement with the size distribution of complex Y-Ti-O oxide particles in Figure 4.24. The influence of Ti on the oxide types will be further discussed in section 5.5.

5.2 Effect of TMP

The influence of TMP on the microstructure and the mechanical properties of ODS steels has been widely investigated [16, 68, 100, 101]. A large characterization program in relation to microstructure and mechanical properties has been performed on a 9% Cr RAFM ODS steel EUROFER ODS under different thermo-mechanical treatments (or thermo-mechanical processing). All EUROFER ODS steels exhibit clearly higher creep strength than RAFM steels with a rupture time of 10,000 hours at 600 °C [16]. The biggest progress by TMP is the pronounced improvement of Charpy impact properties with 40% increase in USE and a shift of DBTT from 60 and 100 to -40 and -80 °C. TEM study clearly revealed that the improved impact toughness can be directly linked to the changes in morphology and size of the $M_{23}C_6$ precipitates [68]. Moreover, TMP introduces elongated grains along rolling direction and evident crystallographic textures [102].

5.2.1 TMP influence on microstructure

As shown in Figure 4.7, coarse Cr carbides are observed in the 13.5Cr hipped ODS steels. These carbides can be easily eliminated by annealing treatment at 1050 °C. A small number of Cr oxides and Ti oxides (in section 4.2.2) exhibit high stability during annealing treatment and TMP. No observable difference is found on the morphology, size and number density of these oxides after TMP. Once Cr- and Ti-oxides are formed, they remain stable and cannot be dissolved well below their melting point, which is 1650 °C for Cr_2O_3 and 1670 °C for TiO_2, respectively.

The 13.5Cr ODS alloy without Ti exhibits equiaxed grains with a size of 1–8 μm and a low dislocation density (Figure 4.8). Ti-containing 13.5Cr ODS steels show a bimodal distribution of coarse grains, a few micrometers in size, and smaller grains, about 200–500 nm in size. The darker contrast from nanoscale grains indicates higher dislocation density (Figure 4.9). This structure was quite stable when it was formed and it is very difficult to be eliminated by short time (2 h) heat treatment at 1050 °C (Figure 4.14) or TMP (Figure 4.18). Since cross rolling technique is employed, coarse elongated grains are observed in both RD-ND section and TD-ND sections. Uniform equiaxed grains, a few microns in diameter, are found in the RD-TD section (Figure 4.17). As pointed out by Oksiuta et.al. [103], the bimodal grain size distribution and heterogeneous dislocation density between the coarse and fine grains are detrimental to the impact property for 14Cr ODS steels. The texture of 13.5Cr1.1W ODS steel after TMP, consists primarily of the α- (<110> // RD) texture fiber which indicates strong recovery (Figure 4.22). The γ- (<111> // ND) texture fiber which is typical of recrystallized rolled ferritic steels does not appear.

Strong recovery is further confirmed by the presence of coarse elongated grains and a large fraction of low angle grain boundaries. ODS alloys tend to recrystallize at temperatures close to their melting point [104]. This unique behavior found in ODS alloys can be attributed to the presence of finely-dispersed Y_2O_3 precipitates that inhibit static recrystallization [105]. Large carbides ($M_{23}C_6$) may act as nucleation sites and promote static recrystallization via particle stimulated nucleation whereas nanoscale oxide particles, particularly if closely spaced, exert a significant pinning effect on both low and high angle grain boundaries. The unusual high recrystallization temperature indicates that nanoscale oxide particles play a decisive role on recrystallization of ODS alloys.

5.2.2 TMP influence on tensile properties

The stability of second-phase precipitates and bimodal grain size distribution lead to constant hardness (Figure 4.47) and tensile properties (Figure 4.51) for HIP+TMP ODS alloy in comparison to HIP ODS alloy. No anisotropy in tensile strength is observed for the HIP+TMP ODS steel with different orientations over the whole temperature range. Remarkable anisotropy is observed in total elongation. Among all ODS materials, the highest total elongation of 21.6% is reached at 500°C for T orientation in HIP+TMP steel. The magnitude of the total elongation obtained for L orientation is about 15–50% lower, but it remains always above 5%. The inferior ductility for 13.5Cr ODS ferritic steel can be explained as follows:

Firstly, γ-fiber, which is helpful for the improvement of the ductility, is absent after TMP. Instead, a sharp α-fiber is observed indicating strong recovery. Dynamic recovery as a softening mechanism is able to reduce the apparent work hardening rate. During tensile deformation, the dislocations that carry the intragranular strain are trapped at grain boundaries and the ODS dispersoids. Park et al. [106] calculated the approximate times for dislocations spreading into grain boundaries. They showed that for ultrafine grained steels (average grain sizes between 1 and 2 µm in diameter), the time for dislocations spreading into grain boundaries is shorter than the deformation time of the tensile test. The grain boundaries and ODS dispersoids significantly reduce the number density of mobile dislocations inside the grains and, consequently, lead to less work hardening.

Secondly, the decrease in tensile ductility can be explained in terms of plastic instabilities, which initiate necking due to excessive localized deformation. The condition for the initiation of necking in a tensile test is indicated by the Conside`re criterion [107]. When the slope of the true-stress true-strain curve (work hardening rate), $d\sigma_t/d\varepsilon_t$, is equal to the true-stress, σ_t, uniform deformation stops and necking is initiated. Oxide dispersion strengthening and grain refinement greatly increases the flow stress of steels, especially during the early stages of plastic deformation. Conversely, they led to a reduced work hardening capacity. Therefore plastic instability (necking) occurs at an early stage during tensile testing, which results in a limited uniform elongation in 13.5Cr1.1W ODS steels after TMP.

5.2.3 TMP influence on Charpy impact properties

In the present work, TMP displays a positive effect on Charpy impact property, with a slight increase in USE from 1.5 J to 2.1 J for 13.5Cr1.1W ODS steels

(Figure 4.56). The USE is much lower than that of base steel and EUROFER ODS steel of 6 J. This may be due to the relatively low ductility of this steel. Another reason can be attributed to the appearance of crack divider type of delaminations in 13.5Cr1.1W ODS steel even in the upper shelf region. Due to limited quantity of ODS steels, Charpy impact tests were only performed with samples in L-T and T-L orientations. The delamination is observed for the samples in both orientations especially tested at room temperature (Figure 4.60).

Delamination is often encountered during Charpy impact tests of hot-rolled high-strength steels, and the density of delaminations usually initially increases with the decreasing testing temperature and then goes through a maximum and decrease [108]. The influence of delamination on USE depends on the fracture type. Two fracture types of delamination are shown in Figure 5.7. The occurrence of crack arrester type of delamination requires relatively weak interfaces or planes normal to the notch orientation of the impact test bar (Figure 5.7b). When the notch is located parallel to the weak planes, the crack divider type of delamination occurs (Figure 5.7c and d). For steels, the weak planes are often the interface between the matrix and second-phase precipitates along the RD.

In the present work, the presence of the crack divider delamination in L-T and T-L samples requires only minor plastic deformation and thus less energy consumption. Therefore a reduced USE is expected for 13.5Cr1.1W ODS compared to the base steel. On the contrary, Kimura et al. [109] developed stronger, tougher steel by combining grain refinement and delamination/splitting of the crack-arrester type. The delamination is thought to relax triaxial stress conditions ahead of the advancing crack front and blunt the crack tip when the crack and crack branching planes intersect, thus leading to high impact energy.

The emergence of the delaminations has not so far been sufficiently explained. From the previous studies [108-112], the delamination resulted from the anisotropic microstructure. Such features as elongated grain shape, certain texture characteristics, decohesion of grain boundaries, segregation of impurities, aligned precipitates could lead to delamination either individually or cooperatively. After TMP, the microstructure of 13.5Cr1.1W ODS steel consists of elongated grains along RD and a sharp (100) <110> α-fiber texture. It is well known that bcc iron cleaves on {100} planes and the coherence length on {100} corresponds to the cleavage crack length [109]. In the elongated grain structure with a strong <110>//RD deformation texture, lots of {100} cleavage planes exist along the RD. The coherence length on {100} is long because of the elongated grain. This structure produces the condition where the crack can run along the longitudinal direction rather than the transverse direction. The aligned precipitates (Figure 4.17) at grain boundaries favor delamination as well.

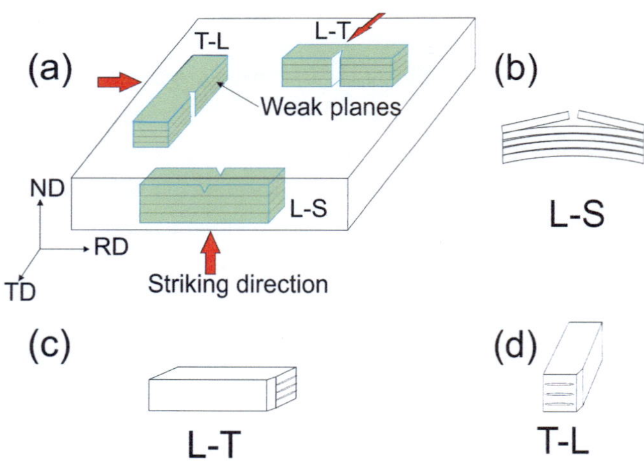

Figure 5.7: Fracture type of delamination / splitting observed after Charpy test.

5.3 Fatigue-microstructure correlation

Along with radiation damage, thermo-fatigue damage is currently considered as the most detrimental lifetime phenomenon for blanket modules. Previously, isothermal tests were performed on RAFM steels at a chosen temperature (often maximum operational temperature for RAFM) [113-115]. The isothermal fatigue data are the basis of fatigue life prediction analysis though this approach is non-conservative in some cases. In order to simulate more accurately the service conditions, Petersen et.al [116, 117] conducted thermo fatigue tests on several RAFM steels, like F82H mod., OPTIFER IV and EUROFER 97. Nevertheless, these experiments were very expensive; they are neither standardized nor applied in design codes. Therefore, isothermal fatigue data is indispensable as the basis and the supporting evidence for lifetime predication simulation.

5.3.1 A comparison of LCF behavior among three ODS steels

ODS ferritic steels are prospective structural materials for the blanket system and first wall components in Tokomak-type fusion reactors. However, both the data and knowledge of fatigue for ODS steels are very limited [52, 118].

The comparison of the isothermal fatigue behavior of EUROFER ODS and 13.5Cr1.1W ODS steels in respect to total strain range vs. number of cycles to failure is depicted in Figure 5.8. Fatigue behavior is high sensitive to a number of variables, including mean stress level, sample geometry, surface effects and fabrication history. The fatigue data in Figure 5.8 for different materials was obtained by samples of the same geometry and surface conditions. After TMP, 13.5Cr1.1W ODS steel exhibits a remarkable lifetime extension irrespective of the strain range. Lifetime prolongation with a factor of 10 to 20 is observed for strain ranges $\Delta\varepsilon \leqslant 0.7\%$. In comparison to EUROFER ODS, fatigue life of 13.5Cr1.W ODS steel after TMP is closely related to the strain range. If the total strain range exceeds around 0.7%, EUROFER ODS steel shows superior fatigue resistance to 13.5Cr1.W ODS steel after TMP. The fatigue life of the 13.5Cr1.1W ODS steel is markedly improved with decreasing total strain range. For example, the sample tested at $\Delta\varepsilon=0.54\%$ remains unbroken after 168,522 cycles.

Figure 5.9 compares the evolution of the positive stress amplitude as a function of fatigue life for different ODS steels. The stress amplitude of 13.5Cr1.1W ODS steel after TMP is typically 450 MPa, two times higher than that of base steel at $\Delta\varepsilon=0.7\%$. The hipped 13.5Cr1.1W ODS steel shows higher stress amplitude, about 480 MPa, along with pronounced lifetime reduction by a factor of 10. In addition, 13.5Cr ODS steel shows no cyclic softening at all during the whole testing process irrespective of the strain range. Most evident cyclic softening of 100 MPa and 70 MPa occurs for 9Cr EUROFER and EUROFER ODS, respectively.

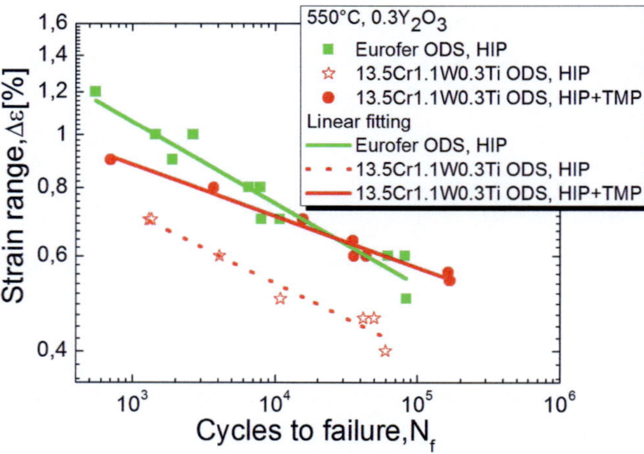

Figure 5.8: Comparison of fatigue properties for different ODS steels as a function of cycles.

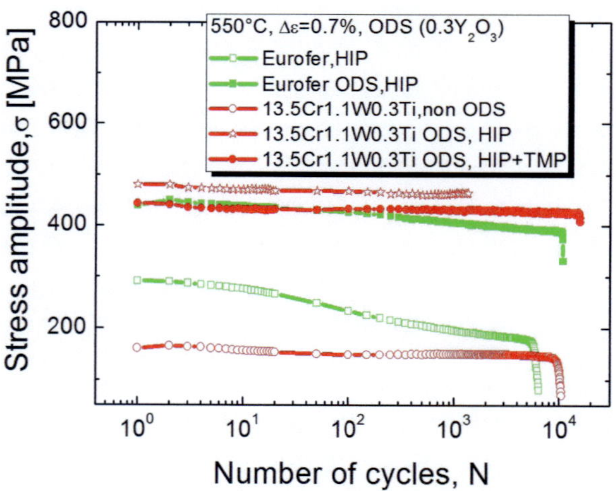

Figure 5.9: Stress amplitude of different ODS steels in comparison to corresponding base steel at Δε=0.7% at 550 °C.

5.3.2 On the microstructure–fatigue relationship in ODS steels

The microstructural mechanisms responsible for this evident cyclic softening of EUROFER have been identified by Vorpahl et.al [119]. TEM micrographs shown in Figure 5.10 reveal the microstructure evolution of EUROFER prior to and after LCF fatigue test. The reduction of dislocation density, the formation of subgrain-structures as well as the formation of precipitates at the grain boundaries and their coarsening during the fatigue stage have been considered as the main contributions for the pronounced cyclic softening of the material.

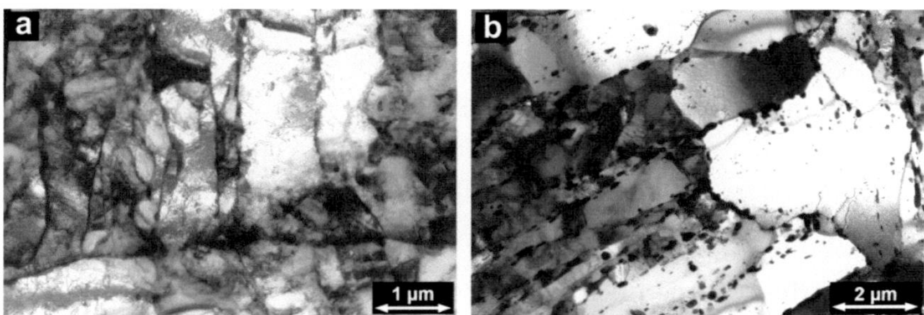

Figure 5.10: Microstructure (TEM) for EUROFER steel in the as-received condition (a), and after pure fatigue at Δε = 0.45% at 550 °C for 60,850 cycles (b) [119].

In case of 13.5Cr1.1W ODS steel, the microstructural evolution was investigated on the sample tested at Δε=0.9% by TEM. The fatigue-induced microstructural evolution will be analyzed in terms of grain structure, dislocation density and chemical segregation.

Prior to fatigue test, 13.5Cr1.1W ODS steel features fine elongated grain along RD with a sharp α-fiber texture and a large fraction of low angle grain boundaries (LAGB), as discussed in 5.2.1. Moreover, equiaxed grains are also found with a broad size distribution. Equiaxed grains less than 500 nm in diameter show the highest residual strain. Coarse equiaxed grains are an indication of primary recrystallization with a dislocation density within the grain. As shown in Figure 4.68, the elongated grains and nanoscale equiaxed grains are still visible after fatigue test. Low angle grain boundaries are often found among these elongated grains. It should be noted that EBSD orientation map and TEM micrographs display slightly different grain size for the same material. This difference stems from different means of identifying grains and grain boundaries utilized in the corresponding techniques [120]. TEM is capable of capturing small misorientations, and returns an intermediate grain size. Nevertheless, limited area of TEM investigation is an important restriction for accurate quantitation analysis, especially for the heterogeneous material. Compared to TEM, EBSD shows larger grain size because of large scanning area and step size. Furthermore, EBSD provides invaluable information that is not easily obtainable from TEM, namely the respective volume fractions of LAGBs and HAGBs. Due to the limited volume of fatigue samples, microstructure characterization after fatigue test is only performed by TEM. Although TEM micrographs in Figure 4.68 reveal the presence of LAGBs, the accurate determination of the respective fraction of LAGBs and HAGBs is very difficult. Therefore, there is no direct evidence for the formation of sub-grain structure during LCF.

However, the formation of sub-grains leads to a dramatic drop of the free dislocation density in the grain interior. Although no quantitative calculations with respect to dislocation densities were conducted for EUROFER, a sharp reduction in dislocation density is clearly visible from high-quality TEM micrographs (Figure 5.10). In order to make more accurate analysis, the dislocation density for 13.5Cr1.1W ODS steel is presented by the mean value of 3 individual measurements at different locations (Table 4.13). The calculations reveal constant dislocation densities around 10^{-14} m$^{-2,}$ independent of the number of cycles. The constant dislocation density suggests the stable grain structure during fatigue test.

Random distribution of a small density of spherical Ti-enriched precipitates is observed in as-received sample as well as in the sample after fatigue test. No evident chemical segregation at grain boundaries occurs during fatigue test for 13.5Cr1.1W ODS steel compared to EUROFER.

5.3.3 Origin of the structural stability

It has been known for a decade that the extraordinary mechanical properties of ODS steels originate from highly stabilized nanoscale oxide. The complex microstructure of ODS steels offers a combination of several strengthening mechanisms. The strengthening resulting from oxide dispersion is roughly estimated by the difference of tensile strength between ODS steel and the corresponding base steel. Figure 5.11 compares tensile strength for EUROFER ODS and 13.5Cr1.1W ODS steel. The difference of the tensile strength between two ODS steels varies slightly with the testing temperature whereas EUROFER is much stronger than 13.5Cr1.1W0.3Ti base steel. The oxide dispersion strengthening leads to an increase in tensile strength by a factor of 160% for 13.5Cr1.1W0.3Ti ODS steel and 38% for EUROFER ODS, respectively. The huge variation in oxide dispersion strengthening of these two ODS steel could be attributed to the pre-alloyed 0.3% Ti. The addition of 0.3% Ti shows striking influence on the particle refinement by shifting the average diameter to 3.73 nm (Figure 4.25). A wide range of particle sizes is observed for EUROFER ODS, with a maximum around 12 nm. Nevertheless, even particles up to 40 nm can be found as well [121]. Moreover, a dramatic increase in number density and a decrease in inter-particle spacing are accompanied by the refinement of oxide particles (Figure 4.31).

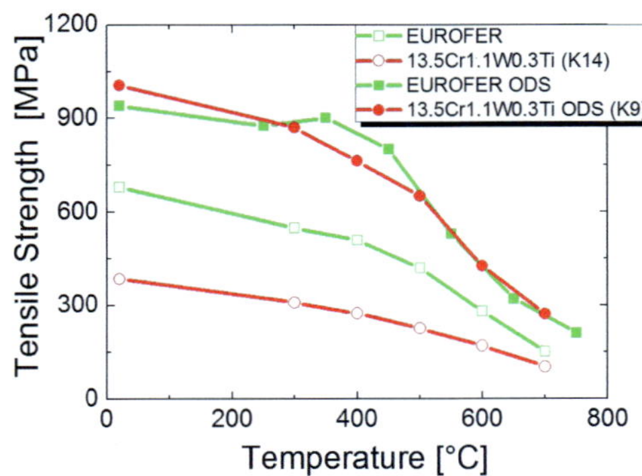

Figure 5.11: Comparison of tensile strength for two ODS steels [31].

5.3.4 Strain-dependent fatigue process

As shown in Figure 5.8, the fatigue lifetime between EUROFER ODS and 13.5Cr1.1W ODS steel is largely dependent on the applied strain range. The discussion of the fatigue process will be classified into two categories: low $\Delta\varepsilon_t$ and high $\Delta\varepsilon_t$.

Low $\Delta\varepsilon_t$ (0.54%)

Basically, the fatigue life comprises two periods, the crack initiation period and the crack growth period. In the crack initiation period, fatigue is a material surface phenomenon. The origin of fatigue cracks in metals and alloys is caused by the irreversible dislocation slip and the subsequent surface roughening. This roughning is manifested as microscopic extrusions and intrusions. The intrusions function as micronotches and the effect of stress concentration at the root of the intrusions promotes additional slip and fatigue crack nucleation. At low $\Delta\varepsilon_t$, the crack initiation period covers a relatively large percentage of the total fatigue life. Microstructurally small cracks can be nucleated at very low stress amplitudes. After 168,522 cycles at $\Delta\varepsilon$=0.54%, a high density of micro crack segments are clearly visible on the lateral surface of 13.5Cr1.1W ODS steel. The microcracks propagate preferably by the coalescence of these micropores. In most cases, these crack segments align approximately parallel to the main loading direction (the same with RD). This is inconsistent with that observed in F82H with a maximum distribution between 40° and 50° [113], but can be explained by the elongated grains parallel to RD. The aggregate boundaries are favorable sites for crack initiation and crack propagation.

After isolated microcracks have developed to a characteristic length, damage accumulation has been shown to continue by crack growth. In the crack growth period, fatigue in these ODS steels is depending on the crack growth resistance of the material, rather than on the material surface conditions. At low $\Delta\varepsilon_t$, the crack growth is significantly impeded by very fine oxide particles before they are able to reach barriers like grain boundaries (Figure 4.70). Therefore, evident lifetime prolongation is obtained for 13.5Cr1.1W ODS steel at low $\Delta\varepsilon_t$.

High $\Delta\varepsilon_t$(0.9%)

At high strain ranges microstructural barriers are already surmounted during the first few cycles and thus crack initiation period covers only a small fraction of fatigue life. This is followed by early network formation of cracks, which leads to a reduction in fatigue life by a factor of 2. This is confirmed by only two main cracks along with pronounced less crack segments after fatigue at $\Delta\varepsilon$=0.9%. Two main cracks have an orientation of 45° to 60° relative to the loading direction. This can be interpreted by a superposition of maximum shear stress driven plastic deformation close to 45 ° and interface separation induced by normal stress.

5.4 Quantitative contributions from various strengthening sources

The very high yield strength of 13.5Cr ODS steels result from different strengthening sources, for example, grain refinement and high dislocation density. However, the ultrafine oxide particles with a high number density are considered as the major source of dispersed obstacle hardening. This section attempts to access the individual contributions from various sources by a linear combination according to Equation 2.1. It should be kept in mind that the assumption of linear superposition of these various contributions may not be valid in some cases. This will aggravate the uncertainty in the quantitative analysis. It would be possible to specify the precise contributions of all these mechanisms only if there are complete information on all details of the micro-nano structure and chemistry of the alloys coupled with rigorous physical models, including proper superposition rules. Currently, the full information and sufficiently detailed models are not available. Therefore the quantitative evaluation of various strengthening sources has been conducted in a more empirical and approximate manner.

The lattice friction σ_{lf} and solid solution σ_{ss} at room temperature is represented by the yield strength (261 MPa) of 13.5Cr1.1W0.3Ti base alloy after TMP.

5.4.1 Strain hardening σ_d

The contribution from dislocation hardening is expressed as:

$$\sigma_d \approx 1.2Gb\sqrt{\rho} \tag{5.3}$$

Where G is the shear modulus, b is the Burger's vector ($\approx 2.5 \times 10^{-10}$ m) and ρ is the dislocation density. At room temperature G is 68 GPa by using Young's modulus E of 175 GPa and the linear isotropic elasticity formula:

$$G = E/2(1+\nu) \tag{5.4}$$

Where the Poisson's ratio ν is 0.3 for iron.

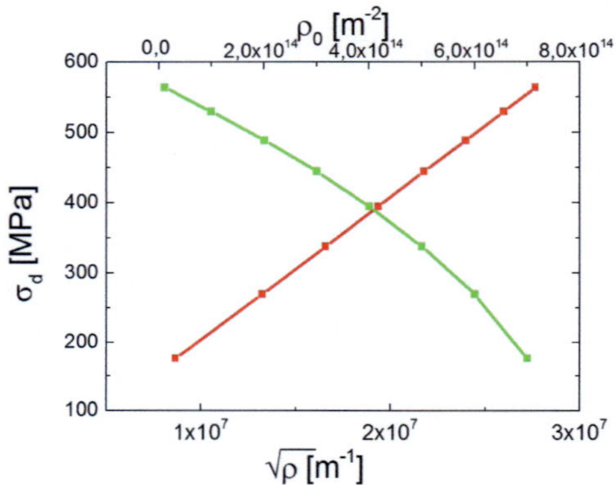

Figure 5.12: Stress increment σ_d by strain hardening as a function of $\sqrt{\rho}$. The estimated dislocation density of the base alloy K13 is also included.

Although the dislocation density in 13.5Cr1.1W0.3Ti base steel has not been measured yet, the hipped base alloy shows a lower dislocation density than that of the ODS variant. TMP leads to no observable influence on the yield strength for 13.5Cr1.1W base steel (Figure 4.51b). This indicates constant dislocation densities prior to and following TMP. The dislocations introduced by MA and hot rolling have been restored to the initial level by fully recrystallization during subsequent annealing. The dislocation densities vary from 10^{13} m^{-2} [54] for fully recrystallized materials to 7.75×10^{14} m^{-2} (Table 4.13) for ODS steel after TMP. Figure 5.12 shows a broad range of σ_d from 170 MPa to 560 MPa due to the uncertainty of ρ_0 for the base alloy K13. An estimation by Equation 5.3 leads to a maximum value of 560 MPa at room temperature when $\rho_0 = 10^{13}$ m^{-2}. A linear relation of σ_d and $\sqrt{\rho}$ is revealed by Equation 5.3 and Figure 5.12.

5.4.2 Grain refinement σ_{gb}

σ_{gb} is related to the grain size by the Hall-Petch relationship Equation 2.2. The grain size of 13.5Cr1.1W base steel and ODS variant is 20 μm and 1.42 μm (Figure 4.20a) on the basis of OM, EBSD and TEM investigation. In order to analyze the strengthening directly, another version of Hall-Petch relation has been employed by Eiselt [64], given as

$$\sigma_{gb} = M\beta Gb \frac{1}{\sqrt{d}} \tag{5.5}$$

Where M is Taylor factor (similar to Equation 2.6), β is constant, d is the average grain size (similar to Equation 2.2), G is shear modulus (similar to Equation 5.4) and b is Burgers vector. In bcc iron, M =2.77–3.06, β =4380 m$^{1/2}$, b = 0.25 nm. With G=68 GPa and d=1.42 μm, σ_{gb} is estimated to be 150–207 MPa.

However, this estimation is not rigorous due to several microstructural factors. Firstly, textural anisotropy could influence the yield stress. However, similar yield stress for L-T and T-L samples suggests little texture effect. Another reason is the grain morphology with a given grain size, for instance, the elongated grains after TMP. It has been revealed [122] that with an identical subgrain size, different misorientation distribution could lead to a variation in yield stress.

5.4.3 Dispersion strengthening σ_p

As stated in section 2.5, only nanoscale oxide particles with much higher number density and volume fraction are considered as the most important contributor for improved high temperature strength. The average particles size was measured as 3–5 nm and the inter-particle spacing is estimated to 30–40 nm.

In 13.5Cr ODS steels, the fine oxide particles can be overcome either by cutting or Orowan by-passing mechanisms. When the shear modulus of precipitates is below the Orowan limit, they can be sheared by the passage of dislocations through them. The critical shear stress necessary for dislocation to pass through a particle, $\Delta\tau_{cut}$, is evaluated by Kubena et al. [52] according to Equation 2.3. Burgers vector b is 0.258 nm and γ is 20 J/m^2 by using atomistic modeling [123].

With a diameter of 3–5 nm and shear modulus G of 68 GPa, $\Delta\tau_{cut}$ is estimated to be 10 GPa in Fe matrix. This value is approximately 30% greater than the maximum theoretical shear strength for iron, computed using ab initio calculations as 7.5 GPa. It is also significantly overestimated in comparison to the yield stress from the tensile test. The possible reason for the overestimation is explained by Takahashi et al. [124]. He found that if a precipitate is sheared by a dislocation, its resistance to dislocation motion is reduced and the overall resistance of the two sheared pieces becomes sufficiently low. The critical shear stress decreases dramatically with an increase of dislocation cuttings. Therefore, many dislocations can slip through the sheared region in an avalanche fashion, and a zone of localized plastic deformation is formed in the matrix.

The critical stress is further rationalized by Orowan by-passing model. The critical stress $\Delta\sigma_o$ originated from complete Orowan mechanism can be described by a general dispersed barrier model as:

$$\Delta\sigma_o = 0.8 M\alpha Gb / \lambda \qquad (5.6)$$

Here M is the Taylor factor ($\approx$ 2.77–3.06), α is a strength factor that may depend on the obstacle size, the $0.8/\lambda$ is an effective spacing of features randomly located on a slip plane. In principle, the α factor is typically referred to as the barrier strength and accounts for the fact that some obstacles may be partially cut or sheared by the dislocation as it bows out. This reduces the amount of energy required for a dislocation to glide through the field of obstacles. For impenetrable Orowan obstacles, α=1.0 whereas for cutting mechanism, α is close to 0. However, very few measurements of α have been made. α=0.25 was proposed by Eiselt [64] to give the best agreement of the prediction and the experimental result. σ_p is estimated to 330 $\pm$ 15 MPa by Equation 5.6 assuming that α=0.25. It can be concluded that the Orowan mechanism of dislocation is energetically favored to the particle cutting process.

Figure 5.13 shows quantitatively different strengthening contributions in current ODS steels K9. It is confirmed that the best fitting is obtained when taking α=0.25. The greatest strengthening effect, approximately 50%, originates from nanoscale oxide particles. Furthermore, comparable contributions are observed from both straining hardening and grain size refinement.

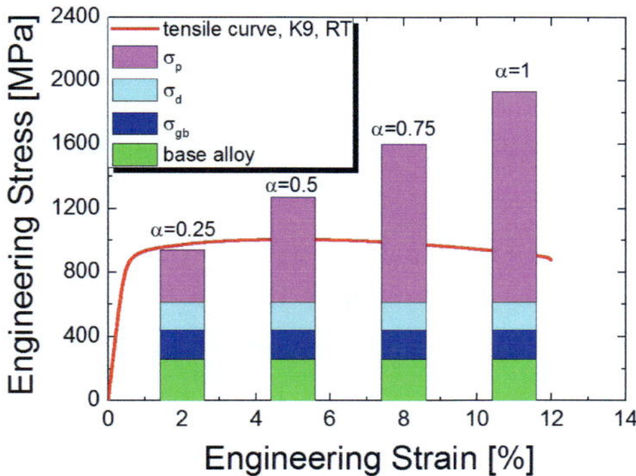

Figure 5.13: The comparison of the experimental tensile result with the prediction according to numerous strengthening models as a function of a strength factor α in Equation 5.6.

5.5 Evaluation of XAFS application

5.5.1 Advantages and limitations of XAFS

Over the past 30 years, XAFS has been successfully applied to a wide range of scientific fields due to the rapid development of synchrotron sources [125, 126]. However, very few XAFS research has been conducted for ODS alloys. The application of the XAFS technique for ODS alloys is evaluated as follows:

(1) XAFS is element specific, so one can focus on one specific element without interference from other elements present in the sample.

(2) High brilliant third generation synchrotron radiation provides the opportunity for the study of dilute samples. XAFS measurements can be made on elements of minor and even trace contents (down to several ppm). The addition of Y_2O_3 in ODS steels is mainly between 0.2% and 0.5% (wt. %). A recent study [23] has successfully demonstrated the primary characterization of precipitations in two ODS steels by a complementary application of XAFS and XRD using synchrotron irradiation sources.

(3) XAFS is an atomic probe and is not limited by the state of sample. Importantly, crystallinity is not required. It is only sensitive to the local structure of absorbing atoms which makes it one of the few powerful tools available for

noncrystalline and highly disordered materials. This provides a perfect tool for MA powders with very high dislocation densities and disordering.

(4) There are no strict requirements for the shape and the size of the specimen. Compared with TEM, no vacuum is required, specimens are easy to make and no artifacts are introduced during sample preparation. A large beam size allows the measurement of a large illuminated area so that the adverse influence of microstructural heterogeneity in ODS alloys can be eliminated.

(5) XAFS is a non-destructive method.

However, the limitations are

(1) Though the measurement of XAFS is simple, a comprehensive understanding of the results involves a combination of modern physics, chemistry, materials science, and a complete mastery of data analysis.

(2) XAFS provides only indirect evidence of speciation of oxide phases. The quantitative evaluation of XAFS by using data fitting also depends largely on the availability of the references. For Y K-edge, only yttrium metal and Y_2O_3 are investigated. Other complex oxides, for instance, Y_2TiO_5, $Y_2Ti_2O_7$, which are often present in the compacted ODS steels, are currently not commercially available. High disordered reference is essential for the characterization of MA powder as well. Therefore, the absence of essential references results in the difficulties for the accurate phase identification. Other complementary techniques such as TEM or XRD are always indispensable to make the correct determination.

(3) In principle, many experimental techniques and sample conditions are available for XAFS. In practice, the characteristics of synchrotron sources and experimental station including the energy ranges, beam sizes, and intensities, often put practical experimental limits on the XAFS measurement. For example, the strong dependence of μ on Z^4 and E is a fundamental property. Due to the Z^4 dependence, reasonable absorption coefficient cannot be achieved for materials with very low Z. Besides, the energy for XAFS ranges roughly from 3 to 40 keV. Noisy data is often obtained for the materials with absorption edge E_0 close to the low and high energy restriction. Specifically, poor data for Ti K-edge

(E_0= 4.966 keV) were obtained in comparison to that for Y K-edge (E_0= 17.038 keV) and very limited information can be deduced on the basis of noisy data.

(4) Very long cycle and relative poor availability of synchrotron radiation beamtime is an important restraint.

5.5.2 Evolution of Y_2O_3

MA is a critical step to study the formation mechanism of nanoscale second-phase oxides. Numerous variables such as type of the mill [33], milling time [61, 63], mill speed and milling atmosphere [62] have been extensively investigated. The mean particle size extracted from XRD data has been used to determine the optimum milling time in previous studies [61, 63]. A new approach [127] for TEM sample preparation of the MA powders has been already demonstrated recently. It was found that the dissolution of Y_2O_3 is strongly dependent on the milling time. Complete dissolution of Y_2O_3 is observed after 80 hours milling with a speed of 1200 rpm. In this work, XAFS has revealed that only 10–14% of initially added 0.3% Y_2O_3 dissolved into the steel matrix after 24 hours milling with the same milling machine and ball-to-powder ratio. It suggests that a long-term milling is required for the complete dissolution of Y_2O_3. Different from the initial Y_2O_3 (Figure 4.41), the undissolved Y_2O_3 shows highly disordered structures. The energy introduced by MA destructs the long range ordering of α-Y_2O_3. However, this energy is insufficient to break the ultra-stable Y-O bonds.

There is considerable controversy over the structure of the oxide particles in compacted ODS alloys. A wide variety of particle structures has been reported in various alloys. TEM analyses showed that some particles larger than 15 nm were consistent with the $Y_2Ti_2O_7$ phase in MA957 [42]. SANS data obtained from the 14YWT alloy [33] and TEM analysis on 18% Cr containing steel suggested a mixture of Y_2TiO_5 and $Y_2Ti_2O_7$ [39, 128]. For particles with radii <2 nm, Miller et al. measured a Y: Ti ratio of considerably less than 1 (ranging from 0.25 to 0.5) by APT, with an overall metal to oxygen ratio (M:O) of 1:1 [43-45].

Table 5.1: Enthalpy of formation of oxides (data for 298K) [129].

Oxide phase	M:O ratio	Enthalpy of formation $-\Delta H_f$ per mole of oxide [kJ mol^{-1}]
CrO_2	1:2	583
Cr_2O_3	2:3	1130
TiO_2	1:2	944
Ti_3O_5	1:1.67	2457
Ti_2O_3	2:3	1522
TiO	1:1	543
$YCrO_3$	2:3	1493
$Y_2Ti_2O_7$	1:1.75	3874
Y_2O_3	2:3	1907

As shown in Figure 4.43, Y-enriched oxides in compacted ODS steels are more dispersed and complicated than pure Y_2O_3. This suggests the coexistence of crystalline nanoparticles and disordered phases after high temperature consolidation and annealing. On a basic level, the precipitation of the nanoparticles is driven by the low equilibrium solubility of O in the matrix. First and foremost it is an oxidation reaction that causes a nanoparticle embryo to form, driven by the change in free energy associated with that reaction. Table 5.1 shows the formation enthalpies for many of the possible stable oxides. The formation of complexes with Y and Ti is expected given that these elements have a high affinity for oxygen.

For the Ti free ODS alloy, the coexistence of Cr oxides and Y-Cr-O/Y-O particles is confirmed by EDX elemental mapping (Figure 4.33) and the EXAFS data fitting (Figure 4.44). This is consistent with the particles having structures of Y_2O_3, $YCrO_3$ and Cr_2O_3 in a Ti free model alloy [130]. The EDX mapping in Figure 4.28 indicates the presence of Y-Ti-O particles in 0.4Ti ODS alloy. Amongst these known stable oxides, the formation of $Y_2Ti_2O_7$ is considered most likely. Due to the incomplete dissolution of Y_2O_3 during MA, the existence of Y_2O_3 in Ti-containing alloy cannot be excluded.

The kinetics of the nanoparticle nucleation will depend on the concentrations and diffusivities of the oxide-forming species, and the energy barrier for nucleation. The nucleation barrier can be considered as a combination of a strain energy term corresponding to the change in volume free energy associated with the nanoparticle, and also the extra energy associated with the formation of a metal-oxide interface [129]. In highly deformed alloys, second phase particles are assumed to nucleate heterogeneously on dislocations, and there will be an abundance of such nucleation sites introduced by the MA. Hirata et al. have shown that a defective rock-salt structure can be fully coherent with the matrix, and suggested that this gives rise to a very low interface energy [35]. This is likely to encourage the formation of non-stoichiometric nanoparticles over the more thermodynamically stable $Y_2Ti_2O_7$. Recently, both crystalline nanoparticles larger than 2 nm and amorphous (or disordered) cluster domains smaller than 2 nm have been observed in MA956 and Fe-16Cr-4.5Al ODS steels by HRTEM [49]. This could also explain the dispersed oxide phases observed in Figure 4.43 and poor data fitting with any known Y-O or Y-Ti-O phases for the Ti-containing ODS samples.

6 Summary and outlook

6.1 Summary

ODS steels possess excellent high temperature strength, stability and expansion resistance and have been extensively developed as a promising material for fission and fusion reactor applications. The objective of this work is to improve mechanical properties by TMP. The mechanical properties for ODS steels with different preparation parameters were evaluated with the aid of mechanical experiments. The microstructure and the nanoscale oxide particles were characterized in different processing states by a set of complementary techniques to provide key insights into microstructure-property correlation. Furthermore, the combined use of XAFS and TEM provide unique opportunities for a greater understanding of the formation mechanism of the nanoparticles. The main conclusions in relation to specific goals derived from the current research and the literature are summarized in the following:

(1) The influence of chemical composition

In order to achieve higher plant operation temperature without losing the merits inherent in RAFM steels, worldwide efforts have been devoted to develop fully ODS ferritic steels. 12% Cr is often chosen to ensure a fully ferritic microstructure without the formation of brittle Cr rich phases, for instance, σ FeCr phase and Cr rich α' phase. However, phase transformation has been measured by the dilatometry which is further confirmed by ferritic/martensitic structure for 12% Cr alloys (K15, K7, K8). This can be accounted for the high Ni concentration (1.6%) in the 12% Cr base powder K15. The evident variations in M_s and M_f for three 12Cr2W steels (K15, K7, K8) indicate that the addition of Y_2O_3 and Ti have profound influence on the kinetics of martensitic transformation.

Slightly higher Cr content of 13.5% is chosen to improve the corrosion resistance with fully ferritic structure. The microstructural observation reveals fully ferritic structure in 13.5Cr1.1W0.3Ti alloys and 13.5Cr2W alloys. The grain boundaries for 13.5Cr2W ODS steels (K2, K4, K5, K6) and 13.5Cr1.1W0.3Ti ODS steel (K1) are decorated with elongated Cr rich carbides after HIP. The morphology and the analytical results by means of EDX and EELS are in good agreement with $(Fe,Cr)_{23}C_6$ type observed in "as-HIPped" EUROFER ODS steel. The presence of the Cr carbides in the as-HIPped ODS ferritic steels can be accounted for a slow cooling rate after HIP and thus for an accelerated diffusion rate. These carbides can be easily removed by high temperature annealing at 1050 °C and fast vacuum cooling.

Addition of 1–2% W improves high temperature strength without introducing brittle Laves phase and/or μ phase for 12–13.5%Cr ODS steels. 13.5Cr2W0.3Ti ODS steel K5 exhibits an evident increase in tensile strength between 300 °C and 500 °C in comparison to 13.5Cr1.1W0.3Ti ODS steel K1.

After HIP, the highest hardness is observed for 12Cr ODS steel (K8) due to the presence of ferritic/martensitic structure. The tempering treatment at 750 °C for 2 hours leads to a remarkable decrease in hardness for 12Cr ODS steels. In contrast, hardness increases for 13.5Cr2W and 13.5Cr1.1W0.3Ti ODS steels after the annealing treatment because of the dissolution of Cr carbides and the enhanced solid solution.

(2) The influence of TMP

TMP leads to heterogeneous microstructure in terms of elongated grain morphology and broad grain size range. A sharp α- (<110> // RD) texture fiber introduced by TMP indicates strong recovery. The strong recovery is further confirmed by the presence of a high fraction of low angle grain boundaries. Since recovery and recrystallization are competitive mechanisms, the γ- (<111> //ND) texture fiber which is typical of recrystallized rolled ferritic steels does not appear. In some cases, coarse precipitates are aligned favorably at the grain boundaries parallel to RD.

No anisotropy in tensile strength is observed for the HIP+TMP ODS steel with different orientations over the whole temperature range. More pronounced anisotropy is observed in total elongation. The magnitude of the total elongation obtained for L orientation is about 15–50% lower, but it remains always above 5%. The inferior ductility for 13.5Cr ODS ferritic steel can be explained by the heterogeneous grain structure, the absence of γ- (<111> //ND) texture fiber, brittle precipitates at grain boundaries and the dispersion of nano oxide particles.

TMP displays a positive effect on Charpy impact property, with a slight increase in USE from 1.5 J to 2.1 J for 13.5Cr1.1W ODS steels. However, this is much lower than that of base steel and EUROFER ODS steel of 6 J. The relative low ductility and the appearance of crack divider type of delaminations in the samples with L-T and T-L orientation are considered as the main reasons for dissatisfactory Charpy impact property. It is likely that the samples with L-S or T-S orientation exhibits much better impact property due to delamination toughening in form of the crack-arrester type.

(3) Fatigue-microstructure correlation

After TMP, 13.5Cr1.1W ODS steel exhibits a remarkable lifetime extension irrespective of the strain range in comparison to the hipped ODS steel. The lifetime prolongation with a factor of 10 to 20 is observed for strain ranges $\Delta\varepsilon$ $\leqslant$0.7%.

Compared to EUROFER ODS, fatigue life of 13.5Cr1.W ODS steel after TMP is closely related to the strain range. If the total strain range exceeds around 0.7%, EUROFER ODS steel shows superior fatigue resistance. The fatigue life of the 13.5Cr1.1W ODS steel is markedly improved with decreasing total strain range and longer lifetime is achieved at low strain range.

The evident cyclic softening for EUROFER can be attributed to the reduction of dislocation density, the formation of subgrain-structures as well as the formation of precipitates at the grain boundaries and their coarsening during the fatigue stage. The constant stress amplitude for 13.5Cr1.1W ODS steel irrespective of strain range is derived from stable grain structure, constant dislocation densities of 10^{14} m^{-2} and no evident chemical segregation at grain boundaries. The extraordinary fatigue resistance of ODS steels originates from highly stabilized nanoscale oxide with an average diameter about 4 nm.

(4) The characteristics of second-phase precipitates

Nano particles

The addition of an appropriate amount of Ti is not only beneficial to refine the particle size and increase the particle number density, it can lead to significant change in the chemical composition of ODS dispersoids. The addition of 0.3% Ti is a very effective approach for the refinement of nanoscale ODS particles by shifting the average diameter to 3.73 nm. Moreover, Y_2O_3 is replaced by Y-Ti-O particles in all Ti-containing ODS alloys. However, the existence of Y_2O_3 cannot be ruled out.

EDX mapping provides reliable proof of the coexistence of Y_2O_3 and a small number of Y-Cr-O particles in Ti free ODS alloy. Non-negligible content of Al is occasionally detected in the nano particles in all ODS alloys. It is believed that Al is picked up during MA by cross contamination.

No shell-core structure is found in any of the analyzed nanoparticles. The very thin shell of 0.5 to 1.5 nm can only be observed with very good signal-to-noise ratio.

Coarse precipitates

Elongated Cr carbides aligned preferentially along grain boundary in all as-hipped 13.5Cr ODS alloys. These carbides may lead to intergranular embrittlement and thus are detrimental to the impact toughness. They are dissolved in the matrix after 2 hours high temperature annealing.

The addition of Ti results in the formation of spherical Ti oxides rather than Cr oxides owing to the stronger affinity of Ti to oxygen.

(5) The formation mechanism of nanoparticles

XAFS provides a non-destructive way to study the physical and chemical state of dilute species in ODS alloys. It enables the study of Y-enriched nanoparticles with a low concentration (usually 0.2–0.5%) by measuring Y K-edge XAFS. The XAFS reveals the average structural information and the dominant species, which is complementary to the knowledge of the individual particles obtained by TEM.

A linear combination fit of the XANES enables a quantitative determination of $Y_{dissolved}/Y_{total}$ for milled powders. After 24 h of milling, 10–14% of the initially added 0.3% Y_2O_3 dissolves into the steel matrix. The varied Ti contents exhibit a minor influence on Y solid solution during the MA. Longer milling time is expected to lead to complete dissolution of Y_2O_3 into the matrix.

After HIP and annealing, similar XANES spectra have been obtained for 0–0.4% Ti ODS alloys. The difference between the initial Y_2O_3 and the compacted ODS alloys at the Y K-edge suggests the formation of new nano phases during the HIP.

Fitting of the EXAFS indicates the coexistence of Y_2O_3 and $YCrO_3$ in the Ti-free compacted ODS sample. EDX elemental mapping validates the presence of complex oxides Y-Cr-O/Cr-O and shows local Fe deficiency at the precipitates.

The more dispersed Y-enriched oxides in the compacted ODS steels indicate the coexistence of crystalline nanoparticles and disordered phases after high temperature consolidation and annealing.

6.2 Outlook

While this work represents an enormous advance in understanding the structure-property correlation and the formation mechanism of oxide paarticles, in many ways it barely scratched the surface of what can be done.

High priorities range from XAFS measurements of samples throughout the various steps of processing over a wide range of alloy chemistries and processing conditions. In practice, the MA time and the annealing temperature are of great significance. Advanced TEM techniques and APT are essential to provide the commentary information in relation to the chemical natures of individual particles or nanoclusters. SANS is also needed to characterize other features of the nanoparticles or nanoclusters including the size, number density and volume fraction. The comprehensive characterization of nanoparticles enables more accurate quantitative evaluation of various strengthening sources.

From a practical point of view, it will be critical to resolve the critical issue of heterogeneity in the nano-microstructures. This should include studies to eliminate bimodal grain size distribution, anisotropic/textured microstructure after TMP. Further, it will be important to achieve a good balance of strength and fracture toughness on the basis of optimizing the nanoparticles and clean steel fabrication. Finally, a large database on the fatigue property of various ODS steels is of great interest for the future fusion application.

In spite of the many challenges that await further development of this very exciting new class of alloys, the current study has laid a very solid foundation for future research.

Bibliography

[1] El-Guebaly LA. Fifty Years of Magnetic Fusion Research (1958-2008): Brief Historical Overview and Discussion of Future Trends, Energies 2010;3:1067.

[2] www.iter.org.

[3] Norajitra P, Bühler L, Fischer U, Gordeev S, Malang S, Reimann G. Conceptual design of the dual-coolant blanket in the frame of the EU power plant conceptual study, Fusion Eng. Des. 2003;69:669.

[4] Norajitra P, Abdel-Khalik SI, Giancarli LM, Ihli T, Janeschitz G, Malang S, Mazul IV, Sardain P. Divertor conceptual designs for a fusion power plant, Fusion Eng. Des. 2008;83:893.

[5] Norajitra P, Giniyatulin R, Hirai T, Krauss W, Kuznetsov V, Mazul I, Ovchinnikov I, Reiser J, Ritz G, Ritzhaupt-Kleissl HJ, Widak V. Current status of He-cooled divertor development for DEMO, Fusion Eng. Des. 2009;84:1429.

[6] Zinkle SJ. Advanced materials for fusion technology, Fusion Eng. Des. 2005;74:31.

[7] Muroga T, Gasparotto M, Zinkle SJ. Overview of materials research for fusion reactors, Fusion Eng. Des. 2002;61-62:13.

[8] Michael D. Nuclear energy: Status and future limitations, Energy 2012;37:35.

[9] http://www.gen-4.org/Technology/roadmap.htm.

[10] van der Schaaf B, Gelles DS, Jitsukawa S, Kimura A, Klueh RL, Möslang A, Odette GR. Progress and critical issues of reduced activation ferritic/martensitic steel development, J. Nucl. Mater. 2000;283–287, Part 1:52.

[11] Klueh RL, Ehrlich K, Abe F. Ferritic/martensitic steels: promises and problems, J. Nucl. Mater. 1992;191-194:116.

[12] Harries DR, Butterworth GJ, Hishinuma A, Wiffen FW. Evaluation of reduced-activation options for fusion materials development, J. Nucl. Mater. 1992;191–194, Part A:92.

[13] Petti DA, McCarthy KA, Taylor NP, Forty CBA, Forrest RA. Re-evaluation of the use of low activation materials in waste management strategies for fusion, Fusion Eng. Des. 2000;51–52:435.

[14] Klueh RL. Reduced-activation bainitic and martensitic steels for nuclear fusion applications, Curr. Opin. Solid State Mater. Sci. 2004;8:239.

[15] Rieth M, Schirra M, Falkenstein A, Graf P, Heger S, Kempe H, Lindau R, Zimmermann H. EUROFER 97, Tensile, Charpy, Creep and Structural tests, Scientific Report 2003;FZKA-6911.

[16] Lindau R, Möslang A, Rieth M, Klimiankou M, Materna-Morris E, Alamo A, Tavassoli AAF, Cayron C, Lancha AM, Fernandez P, Baluc N, Schäublin R, Diegele E, Filacchioni G, Rensman JW, Schaaf Bvd, Lucon E, Dietz W. Present development status of EUROFER and ODS-EUROFER for application in blanket concepts, Fusion Eng. Des. 2005;75-79:989.

[17] Klueh RL, Vitek JM. Elevated-temperature tensile properties of irradiated 9 Cr-1 MoVNb steel, J. Nucl. Mater. 1985;132:27.

[18] Klueh RL, Vitek JM. Tensile properties of 9Cr-1MoVNb and 12Cr-1MoVW steels irradiated to 23 dpa at 390 to 550 ° C, J. Nucl. Mater. 1991;182:230.

[19] Chen CL, Tatlock GJ, Jones AR. Microstructural evolution in friction stir welding of nanostructured ODS alloys, J. Alloys Compd. 2010;504, Supplement 1:S460.

[20] Odette GR, Alinger MJ, Wirth BD. Recent developments in irradiation-resistant steels. Annual Review of Materials Research, vol. 38. Palo Alto: Annual Reviews, 2008. p.471.

[21] Fischer JJ. Dispersion strengthened ferritic alloy for use in liquid-metal fast breeder reactors (LMFBRS). Patent File Date: Filed date 5 Feb 1976, 1978. p.Medium: X; Size: Pages: 10.

[22] Ukai S, Fujiwara M. Perspective of ODS alloys application in nuclear environments, J. Nucl. Mater. 2002;307-311:749.

[23] Kaito T, Ukai S, Ohtsuka S, Narita T. Development of ODS Ferritic Steel Cladding for the Advanced Fast Reactor Fuels. Proceedings of GLOBAL 2005. Tsukuba, Japan, 2005.

[24] Ukai S, Nishida T, Okuda T, Yoshitake T. R&D of oxide dispersion strengthened ferritic martensitic steels for FBR, J. Nucl. Mater. 1998;258–263, Part 2:1745.

[25] Ukai S, Harada M, Okada H, Inoue M, Nomura S, Shikakura S, Nishida T, Fujiwara M, Asabe K. Tube manufacturing and mechanical properties of oxide dispersion strengthened ferritic steel, J. Nucl. Mater. 1993;204:74.

[26] Cho HS, Kimura A. Corrosion resistance of high-Cr oxide dispersion strengthened ferritic steels in super-critical pressurized water, J. Nucl. Mater. 2007;367-370:1180.

[27] Cho HS, Kimura A, Ukai S, Fujiwara M. Corrosion properties of oxide dispersion strengthened steels in super-critical water environment, J. Nucl. Mater. 2004;329-333:387.

[28] Yoshida E, Kato S. Sodium compatibility of ODS steel at elevated temperature, J. Nucl. Mater. 2004;329-333:1393.

[29] Takaya S, Furukawa T, Aoto K, Müller G, Weisenburger A, Heinzel A, Inoue M, Okuda T, Abe F, Ohnuki S, Fujisawa T, Kimura A. Corrosion behavior of Al-alloying high Cr-ODS steels in lead-bismuth eutectic, J. Nucl. Mater. 2009;386-388:507.

[30] Kimura A, Kasada R, Iwata N, Kishimoto H, Zhang CH, Isselin J, Dou P, Lee JH, Muthukumar N, Okuda T, Inoue M, Ukai S, Ohnuki S, Fujisawa T, Abe TF. Development of Al added high-Cr ODS steels for fuel cladding of next generation nuclear systems, J. Nucl. Mater. 2011;417:176.

[31] Lindau R, Möslang A, Schirra M, Schlossmacher P, Klimenkov M. Mechanical and microstructural properties of a hipped RAFM ODS-steel, J. Nucl. Mater. 2002;307-311:769.

[32] Baluc N, Boutard JL, Dudarev SL, Rieth M, Correia JB, Fournier B, Henry J, Legendre F, Leguey T, Lewandowska M, Lindau R, Marquis E, Muñoz A, Radiguet B, Oksiuta Z. Review on the EFDA work programme on nano-structured ODS RAF steels, J. Nucl. Mater. 2011;417:149.

[33] Alinger MJ, Odette GR, Hoelzer DT. On the role of alloy composition and processing parameters in nanocluster formation and dispersion strengthening in nanostuctured ferritic alloys, Acta Mater. 2009;57:392.

[34] Klueh RL, Shingledecker JP, Swindeman RW, Hoelzer DT. Oxide dispersion-strengthened steels: A comparison of some commercial and experimental alloys, J. Nucl. Mater. 2005;341:103.

[35] Hirata A, Fujita T, Wen YR, Schneibel JH, Liu CT, Chen MW. Atomic structure of nanoclusters in oxide-dispersion-strengthened steels, Nat Mater 2011;10:922.

[36] Zinkle SJ, Ice GE, Miller MK, Pennycook SJ, Wang XL. Advances in microstructural characterization, J. Nucl. Mater. 2009;386–388:8.

[37] Ukai S, Harada M, Okada H, Inoue M, Nomura S, Shikakura S, Asabe K, Nishida T, Fujiwara M. Alloying design of oxide dispersion strengthened ferritic steel for long life FBRs core materials, J. Nucl. Mater. 1993;204:65.

[38] Klimiankou M, Lindau R, Möslang A. HRTEM Study of yttrium oxide particles in ODS steels for fusion reactor application, J. Cryst. Growth 2003;249:381.

[39] Klimiankou M, Lindau R, Möslang A. Energy-filtered TEM imaging and EELS study of ODS particles and Argon-filled cavities in ferritic-martensitic steels, Micron 2005;36:1.

[40] Klimenkov M, Lindau R, Möslang A. TEM study of internal oxidation in an ODS-Eurofer alloy, J. Nucl. Mater. 2009;386-388:557.

[41] Klimenkov M, Lindau R, Möslang A. New insights into the structure of ODS particles in the ODS-Eurofer alloy, J. Nucl. Mater. 2009;386-388:553.

[42] Sakasegawa H, Chaffron L, Legendre F, Boulanger L, Cozzika T, Brocq M, de Carlan Y. Correlation between chemical composition and size of very small oxide particles in the MA957 ODS ferritic alloy, J. Nucl. Mater. 2009;384:115.

[43] Miller MK, Hoelzer DT, Kenik EA, Russell KF. Stability of ferritic MA/ODS alloys at high temperatures, Intermetallics 2005;13:387.

[44] Miller MK, Hoelzer DT, Kenik EA, Russell KF. Nanometer scale precipitation in ferritic MA/ODS alloy MA957, J. Nucl. Mater. 2004;329-333:338.

[45] Miller MK, Russell KF, Hoelzer DT. Characterization of precipitates in MA/ODS ferritic alloys, J. Nucl. Mater. 2006;351:261.

[46] Wu Y, Haney EM, Cunningham NJ, Odette GR. Transmission electron microscopy characterization of the nanofeatures in nanostructured ferritic alloy MA957, Acta Mater. 2012;60:3456.

[47] Williams CA, Marquis EA, Cerezo A, Smith GDW. Nanoscale characterisation of ODS–Eurofer 97 steel: An atom-probe tomography study, J. Nucl. Mater. 2010;400:37.

[48] Sakasegawa H, Legendre F, Boulanger L, Brocq M, Chaffron L, Cozzika T, Malaplate J, Henry J, de Carlan Y. Stability of non-stoichiometric clusters in the MA957 ODS ferrtic alloy, J. Nucl. Mater. 2011;417:229.

[49] Hsiung LL, Fluss MJ, Tumey SJ, Choi BW, Serruys Y, Willaime F, Kimura A. Formation mechanism and the role of nanoparticles in Fe-Cr ODS steels developed for radiation tolerance, Phys. Rev. B: Condens. Matter 2010;82:184103.

[50] Moeslang A, Adelhelm C, Heidinger R. Innovative materials for energy technology, Int. J. Mater. Res. 2008;99:1045.

[51] Brocq M, Radiguet B, Poissonnet S, Cuvilly F, Pareige P, Legendre F. Nanoscale characterization and formation mechanism of nanoclusters in an ODS steel elaborated by reactive-inspired ball-milling and annealing, J. Nucl. Mater. 2010.

[52] Kubena I, Fournier B, Kruml T. Effect of microstructure on low cycle fatigue properties of ODS steels, J. Nucl. Mater. 2012;424:101.

[53] Malaplate J, Mompiou F, Béchade JL, Van Den Berghe T, Ratti M. Creep behavior of ODS materials: A study of dislocations/precipitates interactions, J. Nucl. Mater. 2011;417:205.

[54] Alinger MJ. On the Formation and Stability of Nanometer Scale Precipitates in Ferritic Alloys during Processing and High Temperature Service. ISBN: 9780496034932. Santa Barbara: University of California, 2004.

[55] Bhadeshia HKDH, Honeycombe RWK. Steels: Microstructure And Properties: Butterworth-Heinemann, 2006.

[56] Zhao M, Li JC, Jiang Q. Hall–Petch relationship in nanometer size range, J. Alloys Compd. 2003;361:160.

[57] Carlton CE, Ferreira PJ. What is behind the inverse Hall–Petch effect in nanocrystalline materials?, Acta Mater. 2007;55:3749.

[58] Humphreys FJ. Recrystallization and related annealing phenomena. New York: Elsevier Science, 2004.

[59] Benjamin JS. Dispersion strengthened superalloys by mechanical alloying, Metall. Trans. I 1970:2943.

[60] Benjamin JS. Mechanical alloying, Sci. Amer 1976;234:40.

[61] Iwata NY, Kimura A, Fujiwara M, Kawashima N. Effect of milling on morphological and microstructural properties of powder particles for High-Cr Oxide dispersion strengthened ferritic steels, J. Nucl. Mater. 2007;367-370:191.

[62] Oksiuta Z, Baluc N. Effect of mechanical alloying atmosphere on the microstructure and Charpy impact properties of an ODS ferritic steel, J. Nucl. Mater. 2009;386-388:426.

[63] Ramar A, Oksiuta Z, Baluc N, Schäublin R. Effect of mechanical alloying on the mechanical and microstructural properties of ODS EUROFER 97, Fusion Eng. Des. 2007;82:2543.

[64] Eiselt CC. Eigenschaftsoptimierung der nanoskaligen ferritischen ODS-Legierung 13Cr-1W-0,3Y$_2$O$_3$-0,3TiH$_2$, metallkundliche Charakterisierung und Bestimmung von Struktur-

Eigenschaftskorrelationen. vol. FZKA Report 7524. Karlsruhe: Karlsruhe Institute of Technology, 2010.

[65] Loh NL, Sia KY. An overview of hot isostatic pressing, J. Mater. Process. Technol. 1992;30:45.

[66] Schatt W, Wieters KP, Association EPM. Powder metallurgy: processing and materials: European Powder Metallurgy Association, 1997.

[67] Miao P, Odette GR, Yamamoto T, Alinger M, Hoelzer D, Gragg D. Effects of consolidation temperature, strength and microstructure on fracture toughness of nanostructured ferritic alloys, J. Nucl. Mater. 2007;367-370:208.

[68] Klimiankou M, Lindau R, Möslang A. Direct correlation between morphology of $(Fe,Cr)_{23}C_6$ precipitates and impact behavior of ODS steels, J. Nucl. Mater. 2007;367-370:173.

[69] Wilkinson AJ, Britton TB. Strains, planes, and EBSD in materials science, Mater. Today 2012;15:366.

[70] Sandim MJR, Sandim HRZ, Zaefferer S, Raabe D, Awaji S, Watanabe K. Electron backscatter diffraction study of Nb_3Sn superconducting multifilamentary wire, Scripta Mater. 2010;62:59.

[71] Pai RV, Krishnan K, Dash S, Mukerjee SK, Venugopal V. Synthesis, characterization and thermal expansion of $Cr_2–xTixO_3+\delta$, J. Alloys Compd. 2010;507:267.

[72] Heintze SD, Cavalleri A, Forjanic M, Zellweger G, Rousson V. A comparison of three different methods for the quantification of the in vitro wear of dental materials, Dent. Mater. 2006;22:1051.

[73] Pešička J, Kužel R, Dronhofer A, Eggeler G. The evolution of dislocation density during heat treatment and creep of tempered martensite ferritic steels, Acta Mater. 2003;51:4847.

[74] Williams DB, Carter CB. Transmission Electron Microscopy: A Textbook for Materials Science: Springer, 2009.

[75] Iakoubovskii K, Mitsuishi K, Nakayama Y, Furuya K. Mean free path of inelastic electron scattering in elemental solids and oxides using transmission electron microscopy: Atomic number dependent oscillatory behavior, Phys. Rev. B: Condens. Matter 2008;77:104102.

[76] Bunker G. Introduction to XAFS: A Practical Guide to X-ray Absorption Fine Structure Spectroscopy: Cambridge University Press, 2010.

[77] Koningsberger DC, Prins R. X-Ray Absorption: Principles, Applications, Techniques of EXAFS, SEXAFS and XANES: Wiley, 1988.

[78] Möslang A. Development of creep fatigue specimen and related test technology, IFMIF User Meeting 2000.

[79] Callister WD. Materials Science And Engineering: An Introduction: John Wiley & Sons, 2007.

[80] Suresh S. Fatigue of Materials: Cambridge University Press, 1998.

[81] Swamy V, Seifert HJ, Aldinger F. Thermodynamic properties of Y_2O_3 phases and the yttrium–oxygen phase diagram, J. Alloys Compd. 1998;269:201.

[82] Guo B, Luo Z-P. Particle size effect on the crystal structure of Y_2O_3 particles formed in a flame aerosol process, J. Am. Ceram. Soc. 2008;91:1653.

[83] Raabe DL, K. Textures of ferritic stainless steels, Mater. Sci. Technol. 1993;9:302.

[84] Degueldre C, Conradson S, Hoffelner W. Characterisation of oxide dispersion-strengthened steel by extended X-ray absorption spectroscopy for its use under irradiation, Comput. Mater. Sci. 2005;33:3.

[85] Liu T, Xu H, Chin WS, Yang P, Yong Z, Wee ATS. Local structures of Zn_1-xTMxO (TM = Co, Mn, and Cu) nanoparticles studied by X-ray absorption fine structure spectroscopy and multiple scattering calculations, J. Phys. Chem. C 2008;112:13410.

[86] de Castro V, Marquis EA, Lozano-Perez S, Pareja R, Jenkins ML. Stability of nanoscale secondary phases in an oxide dispersion strengthened Fe-12Cr alloy, Acta Mater. 2011;59:3927.

[87] Kevorkov AMK, V. F.; Munchaev, A. I.; Uyukin, E. M.; Bolotina, N. B.; Chernaya, T. S.; Bagdasarov, Kh. S.; Simonov, V. I. . Y_2O_3 single crystals: Growth, structure, and photoinduced effects, Crystallography Reports 1995;40:23.

[88] Geller S, Wood EA. Crystallographic studies of perovskite-like compounds. I. Rare earth orthoferrites and $YFeO_3$, $YCrO_3$, $YAlO_3$, Acta Crystallogr. 1956;9:563.

[89] du Boulay D, Maslen EN, Streltsov VA, Ishizawa N. A synchrotron X-ray study of the electron density in $YFeO_3$, Acta Crystallogr., Sect. B: Struct. Sci 1995;51:921.

[90] Mumme WG, Wadsley AD. The structure of orthorhombic Y_2TiO_5, an example of mixed seven- and fivefold coordination, Acta Crystallogr., Sect. B: Struct. Sci 1968;24:1327.

[91] Chtoun E. HL, Garnier P., Kiat J.M. X-Rays and neutrons rietveld analysis of the solid solutions $(1-x)A_2Ti_2O_7$-xMgTiO$_3$ (A= Y or Eu), Eur. J. Solid State Inorg. Chem. 1997;34:553.

[92] Katagiri S, Ishizawa N, Marumo F. A new high temperature modification of face-centered cubic Y_2O_3, Powder Diffr. 1993;8:60.

[93] Munstermann E, Franke P, Seifert HJ. Binary systems and ternary systems from C-Cr-Fe to Cr-Fe-W: Thermodynamic properties of inorganic materials compiled by SGTE, subvolume C: ternary steel systems, phase diagrams and phase transition data: Springer, 2012.

[94] Kipelova A, Belyakov A, Kaibyshev R. Laves phase evolution in a modified P911 heat resistant steel during creep at 923 K, Mater. Sci. Eng., A 2012;532:71.

[95] Prat O, Garcia J, Rojas D, Sauthoff G, Inden G. The role of Laves phase on microstructure evolution and creep strength of novel 9%Cr heat resistant steels, Intermetallics 2013;32:362.

[96] Narita T, Ukai S, Ohtsuka S, Inoue M. Effect of tungsten addition on microstructure and high temperature strength of 9CrODS ferritic steel, J. Nucl. Mater. 2011;417:158.

[97] Kim S, Ohtsuka S, Kaito T, Yamashita S, Inoue M, Asayama T, Shobu T. Formation of nano-size oxide particles and δ-ferrite at elevated temperature in 9Cr-ODS steel, J. Nucl. Mater. 2011;417:209.

[98] Yamamoto M, Ukai S, Hayashi S, Kaito T, Ohtsuka S. Formation of residual ferrite in 9Cr-ODS ferritic steels, Mater. Sci. Eng., A 2010;527:4418.

[99] Kim JH, Byun TS, Hoelzer DT. Tensile fracture characteristics of nanostructured ferritic alloy 14YWT, J. Nucl. Mater. 2010;407:143.

[100] Oksiuta Z, Lewandowska M, Kurzydlowski KJ, Baluc N. Influence of hot rolling and high speed hydrostatic extrusion on the microstructure and mechanical properties of an ODS RAF steel, J. Nucl. Mater. 2011;409:86.

[101] Oksiuta Z, Mueller P, Spätig P, Baluc N. Effect of thermo-mechanical treatments on the microstructure and mechanical properties of an ODS ferritic steel, J. Nucl. Mater. 2011;412:221.

[102] Mateus R, Carvalho PA, Nunes D, Alves LC, Franco N, Correia JB, Alves E. Microstructural characterization of the ODS Eurofer 97 EU-batch, Fusion Eng. Des. 2011;86:2386.

[103] Oksiuta Z, Baluc N. Microstructure and Charpy impact properties of 12-14Cr oxide dispersion-strengthened ferritic steels, J. Nucl. Mater. 2008;374:178.

[104] Bhadeshia HKDH. Recrystallisation of practical mechanically alloyed iron–base and nickel–base superalloys, Mater. Sci. Eng., A 1997;223:64.

[105] Renzetti RA, Sandim HRZ, Sandim MJR, Santos AD, Möslang A, Raabe D. Annealing effects on microstructure and coercive field of ferritic–martensitic ODS Eurofer steel, Mater. Sci. Eng., A 2011;528:1442.

[106] Park K-T, Kim Y-S, Lee JG, Shin DH. Thermal stability and mechanical properties of ultrafine grained low carbon steel, Mater. Sci. Eng., A 2000;293:165.

[107] Song R, Ponge D, Raabe D. Mechanical properties of an ultrafine grained C–Mn steel processed by warm deformation and annealing, Acta Mater. 2005;53:4881.

[108] Yan W, Sha W, Zhu L, Wang W, Shan Y-Y, Yang K. Delamination fracture related to tempering in a high-strength low-alloy steel, Metall. Mater. Trans. A 2010;41:159.

[109] Kimura Y, Inoue T, Yin F, Tsuzaki K. Inverse temperature dependence of toughness in an ultrafine grain-structure steel, Science 2008;320:1057.

[110] Song R, Ponge D, Raabe D, Kaspar R. Microstructure and crystallographic texture of an ultrafine grained C–Mn steel and their evolution during warm deformation and annealing, Acta Mater. 2005;53:845.

[111] Chao J, Capdevila-Montes C, González-Carrasco JL. On the delamination of FeCrAl ODS alloys, Mater. Sci. Eng., A 2009;515:190.

[112] Inoue T, Yin F, Kimura Y, Tsuzaki K, Ochiai S. Delamination effect on impact properties of ultrafine-grained low-carbon steel processed by warm caliber rolling, Metall. Mater. Trans. A 2009;41:341.

[113] Bertsch J, Meyer S, Möslang A. Fatigue behavior and development of microcracks in F82H after helium implantation at 200°C, J. Nucl. Mater. 2000;283-287:832.

[114] Lindau R, Möslang A. Low-cycle fatique properties of the helium-implanted 12% Cr steel 1.4914 (MANET), J. Nucl. Mater. 1991;179-181:753.

[115] Lindau R, Möslang A. Fatigue tests on a ferritic-martensitic steel at 420°C: Comparison between in-situ and postirradiation properties, J. Nucl. Mater. 1994;212-215:599.

[116] Petersen C. Thermal fatigue behavior of low activation ferrite–martensite steels, J. Nucl. Mater. 1998;258-263, Part 2:1285.

[117] Petersen C, Rodrian D. Thermo-mechanical fatigue behavior of reduced activation ferrite/martensite stainless steels, Int. J. Fatigue 2008;30:339.

[118] Ukai S, Ohtsuka S. Low cycle fatigue properties of ODS ferritic-martensitic steels at high temperature, J. Nucl. Mater. 2007;367-370:234.

[119] Vorpahl C, Möslang A, Rieth M. Creep-fatigue interaction and related structure property correlations of EUROFER97 steel at 550 °C by decoupling creep and fatigue load, J. Nucl. Mater. 2011, in press.

[120] Niendorf T, Canadinc D, Maier HJ, Karaman I, Yapici GG. Microstructure–mechanical property relationships in ultrafine-grained NbZr, Acta Mater. 2007;55:6596.

[121] Eiselt CC, Klimenkov M, Lindau R, Möslang A, Sandim HRZ, Padilha AF, Raabe D. High-resolution transmission electron microscopy and electron backscatter diffraction in nanoscaled ferritic and ferritic-martensitic oxide dispersion strengthened-steels, J. Nucl. Mater. 2009;385:231.

[122] De Messemaeker J, Verlinden B, Van Humbeeck J. On the strength of boundaries in submicron IF steel, Mater. Lett. 2004;58:3782.

[123] Takahashi A, Chen Z, Ghoniem N, Kioussis N. Atomistic-continuum modeling of dislocation interaction with Y_2O_3 particles in iron, J. Nucl. Mater. 2011;417:1098.

[124] Takahashi A, Ghoniem NM. A computational method for dislocation–precipitate interaction, J. Mechan. Phys. Solids 2008;56:1534.

[125] Corker JM, Evans J, Rummey JM. EXAFS studies of pillared clay catalysts, Mater. Chem. Phys. 1991;29:201.

[126] Di Vece M, van der Eerden AMJ, Grandjean D, Westerwaal RJ, Lohstroh W, Nikitenko SG, Kelly JJ, Koningsberger DC. Structure of the Mg_2Ni switchable mirror: an EXAFS investigation, Mater. Chem. Phys. 2005;91:1.

[127] Hoffmann J, Klimenkov M, Lindau R, Rieth M. TEM study of mechanically alloyed ODS steel powder, J. Nucl. Mater. 2012;428:165.

[128] Kishimoto H, Alinger MJ, Odette GR, Yamamoto T. TEM examination of microstructural evolution during processing of 14CrYWTi nanostructured ferritic alloys, J. Nucl. Mater. 2004;329-333:369.

[129] Williams CA, Unifantowicz P, Baluc N, Smith GDW, Marquis EA. The formation and evolution of oxide particles in oxide-dispersion-strengthened ferritic steels during processing, Acta Mater. 2013;61:2219.

[130] de Castro V, Marquis EA, Lozano-Perez S, Pareja R, Jenkins ML. Stability of nanoscale secondary phases in an oxide dispersion strengthened Fe–12Cr alloy, Acta Mater. 2011;59:3927.

List of Figures

List of Tables

List of publications

Journal papers

[1] He, P., M. Klimenkov, et al., Characterization of precipitates in nano structured 14% Cr ODS alloys for fusion application. Journal of Nuclear Materials, 2012, **428**(1–3): p131-138.

[2] He, P., T. Liu, et al., XAFS and TEM studies of the structural evolution of yttrium-enriched oxides in nanostructured ferritic alloys fabricated by a powder metallurgy process. Materials Chemistry and Physics, 2012, **136**(2–3): p990-998.

[3] He, P., R. Lindau, et al. (2013). The influence of thermomechanical processing on the microstructure and mechanical properties of 13.5Cr ODS steels. Fusion Engineering and Design; 2013; **88**: 2448.

Conference presentations

[1] P. He, M. Klimenkov, R. Lindau, A. Möslang, *Microstructure- property correlation of 13.5%Cr nanostructured ODS steels for fusion application*, DIANA I: 1st International Workshop on DIspersion Strengthened Steels for Advanced Nuclear Applications, April 4-8, 2011, Centre CNRS Paul Langevin, Aussois, France.

[2] P. He, R. Lindau, A. Moeslang, H.R.Z. Sandim, *Influence of the thermomechanical treatment on the mechanical properties of 13.5Cr ODS steels*, 27th Symposium on Fusion Technology, 24–28 September 2012, Liège, Belgium.

[3] P. He et al. (Invited) *Correlation of microstructure and low cycle fatigue properties for 13.5 Cr1.1W0.3Ti ODS steels*, submitted to 16th International Conference on Fusion Reactor Materials (ICFRM-16), 20–26 October 2013, Beijing, China.

Schriftenreihe
des Instituts für Angewandte Materialien

ISSN 2192-9963

Die Bände sind unter www.ksp.kit.edu als PDF frei verfügbar oder
als Druckausgabe bestellbar.